specialtypress

Testbeds, Motherships & Parasites

Astonishing Aircraft From the Golden Age of Flight Test

Frederick A. Johnsen

specialty press

Specialty Press
838 Lake Street South
Forest Lake, MN 55025
Phone: 651-277-1400 or 800-895-4585
Fax: 651-277-1203
www.specialtypress.com

© 2018 by Frederick A. Johnsen

All rights reserved. No part of this publication may be reproduced or utilized in any form or by any means, electronic or mechanical, including photocopying, recording, or by any information storage and retrieval system, without prior permission from the Publisher. All text, photographs, and artwork are the property of the Author unless otherwise noted or credited.

The information in this work is true and complete to the best of our knowledge. However, all information is presented without any guarantee on the part of the Author or Publisher, who also disclaim any liability incurred in connection with the use of the information and any implied warranties of merchantability or fitness for a particular purpose. Readers are responsible for taking suitable and appropriate safety measures when performing any of the operations or activities described in this work.

All trademarks, trade names, model names and numbers, and other product designations referred to herein are the property of their respective owners and are used solely for identification purposes. This work is a publication of Specialty Press, and has not been licensed, approved, sponsored, or endorsed by any other person or entity. The publisher is not associated with any product, service, or vendor mentioned in this book, and does not endorse the products or services of any vendor mentioned in this book.

Edit by Mike Machat
Layout by Monica Seiberlich

ISBN 978-1-58007-241-0
Item No. SP241

Library of Congress Cataloging-in-Publication Data

Names: Johnsen, Frederick A., author.
Title: Testbeds, motherships & parasites : the greatest aircraft of flight test / Frederick A. Johnsen.
Other titles: Testbeds, motherships, and parasites
Description: Forest Lake, MN : Specialty Press, 2018. | Includes bibliographical references and index.
Identifiers: LCCN 2017029879 | ISBN 9781580072410
Subjects: LCSH: Airplanes–Flight testing–History. | Aeronautics–Research–History.
Classification: LCC TL671.7 .J59 2018 | DDC 629.134/5309–dc23
LC record available at https://lccn.loc.gov/2017029879

Written, edited, and designed in the U.S.A.
Printed in China
10 9 8 7 6 5 4 3 2 1

Front Cover: *Iconic Boeing NB-52B mothership "Balls Eight" departs Edwards AFB with NASA X-38 test vehicle underwing. (NASA via Gerald Balzer Collection)*

Front Flap: *The ultimate mothership for launching a flight test vehicle has to be NASA's Boeing 747 Shuttle Carrier Aircraft (SCA) shown here with orbiter* Enterprise.

Title Page: *Seconds away from launching into space, the North American X-15 is carried to altitude by its Boeing NB-52B mothership with a Northrop T-38 flying chase.*

Table of Contents: *This Wright Typhoon in the nose of a B-17 testbed propelled the bomber while its four original Wright Cyclone radial engines were shut down. (The Boeing Company)*

Back Cover Photos

Top: *Republic's XF-84H "Thunderscreech" was a highly modified turboprop testbed fitted with a supersonic propeller. (USAF via Gerald Balzer Collection)*

Center: *NASA's F-8 Crusader testbed for the supercritical wing design was an aesthetic blend of new aerodynamics and classic Navy jet fighter, as seen in the 1973 photo. (NASA)*

Bottom: *NASA's Boeing 747 mothership launches Shuttle Orbiter* Enterprise *on its Free Flight Four to prove critical approach and landing performance. (NASA)*

DISTRIBUTION BY:

UK and Europe
Crécy Publishing Ltd
1a Ringway Trading Estate
Shadowmoss Road
Manchester M22 5LH England
Tel: 44 161 499 0024
Fax : 44 161 499 0298
www.crecy.co.uk
enquiries@crecy.co.uk

Canada
Login Canada
300 Saulteaux Crescent
Winnipeg, MB R3J 3T2 Canada
Phone 800 665 1148
Fax: 800 665 0103
www.lb.ca

Table of Contents

Acknowledgments 4
Preface 5

Chapter 1: Old School 6
Chapter 2: Parasites with a Purpose 10
Chapter 3: Bomber Bonus Testbeds 32
Chapter 4: Transport Testbeds 69
Chapter 5: Canadian Testbeds 78
Chapter 6: Post-War Motherships 84
Chapter 7: Motherships in the Jet Age 95
Chapter 8: Space Shuttle Carrier Aircraft 109
Chapter 9: Underwing and Useful 117
Chapter 10: Scaled Composites' Motherships 124
Chapter 11: Airframe Mods and Systems Tests 128
Chapter 12: Miscellanea 158
Epilogue 181

Appendix 1: Testbed Aircraft 182
Appendix 2: Mothership Aircraft 185
Appendix 3: Glossary of Acronyms and Abbreviations 186
Appendix 4: The X-15A-2 Report 187

Selected Reading 198
Endnotes 198
Index 201

DEDICATION

For Sharon, who has lived through so many book cycles with me and who does so many things quietly to accommodate that process.

ACKNOWLEDGMENTS

Thanks are due to many, some now deceased, who helped as I studied these unusual aircraft over many years. They include Carl A. Bellinger, Peter M. Bowers, Walter J. Boyne, Jason Chapman, Tony Chong, Tom Cole, John E. Dean, Archie Difante, Jeannine Geiger (Air Force Test Center History Office), Pierre Gillard, Richard P. Hallion, Dennis R. Jenkins, Craig Kaston, Donald Keller, Harvey Lippincott, Michael Lombardi (Boeing Corporate Historian), Museum of Flight (and Dan Hagedorn and Jessica Jones), Stan Piet, Philip Schultz, Ben Sclair (General Aviation News), the Society of Experimental Test Pilots (and especially Paula Smith and Susan Bennett), Kenneth Swartz, R. W. (Bill) Walker, and Gordon S. Williams. Editor Mike Machat and publisher Dave Arnold at Specialty Press are appreciated collaborators in this effort.

And finally, I am reminded of something aviation historian Peter M. Bowers told me about his book-writing process. Pete said he never really finished writing a book; at some point he just had to stop and get it published. So this volume is an earnest effort to bring as many testbeds and motherships together as could be done without turning this into an endless quest with no tangible results. Here are the results; I hope you enjoy them.

Publisher's Note: In reporting history, the images required to tell the tale will vary greatly in quality, especially by modern photographic standards. While some images in this volume are not up to those digital standards, we have included them, as we feel they are an important element in telling the story.

PREFACE

To the cadence of production aircraft design, development, and deployment there exists a complementary network of testbed aircraft and motherships. Aircraft systems, engines, and even aerodynamic components can be tested on an existing airframe before making their debut on a new production model. Motherships can carry research vehicles aloft to extend the reach of the research airframe. And parasite aircraft combinations have capitalized on aspects of the mothership concept to create operational piggyback machines.

Some of the results have been described as nothing less than Frankensteinian; imagine scaled-down B-29 wings and tail surfaces flying on a PT-19 trainer fuselage, or the rakish vertical fin and rudder of the Hughes XF-11 mounted in place of the original rounded rudder on a Hughes-owned Douglas A-20G attack bomber.

Some have been brilliant uses of existing technology to meet an evolving need, such as the vintage wooden Lockheed Vega that served General Electric in the 1960s when the company needed an airframe that would minimize metallic interference for a test program.

Motherships range from World War I–era Curtiss Jennies mounting target gliders above the upper wing center section to the EB-29s and EB-50s that carried the first X-Planes, and the iconic NB-52s that launched the hypersonic X-15 research rocket airplanes on their historic probes of space.

Parasites are defined in this volume as a subset of motherships. They include functional combination aircraft such as the Air Force's Cold War lash-up of a B-36 bomber carrying an RF-84 parasite or World War II Germany's use of explosive-laden twin-engine bombers navigated to a target by a manned single-engine fighter riding on pylons above the unmanned (and releasable) bomber bomb.

When researching this volume, it became apparent that the depth and breadth of testbeds and motherships throughout the history of aviation might never be fully documented. Military and corporate secret-keeping and the rush to test something and move on means that it is likely to be some deserving testbeds, and possibly even motherships, flew under the radar, figuratively at least. So please take this book as an earnest effort to assemble what is known about many of these fascinating and unorthodox aircraft. In addition, know that not every testbed and mothership can be treated within the span of these pages.

The narrative is unabashedly centered on U.S. and Canadian variants, although a number of salient and interesting foreign contributions are included. The sheer volume of testbeds and motherships necessitates limiting this effort to a set of airframes that are noteworthy for things such as their accomplishments, their boldness and ingenuity, and even their spectacular failures. And that selectivity can be an arbitrary process sure to spark lively thought. I hope you enjoy this distilling of the aeronautical testbed and mothership genres.

There's a gray area that bounds the definitions of testbed aircraft and motherships. Some vehicles characterized as motherships in this narrative released missiles or drones that arguably could make the aircraft something other than a mothership. And the term testbed, in this volume, essentially means an aircraft used for proving components not normally associated with that aircraft. Thus, a B-45 carrying a bomb-bay–mounted J47 jet engine is a testbed within the parameters of this book, but an F-86 instrumented to evaluate its own J47 is not considered to be a testbed here. Or else I'd be chasing aircraft for decades just trying to quantify everything that might be called a testbed or a mothership. I hope you find some old favorites, as well as some specialized aircraft that you never encountered before, in this book.

Aircraft alphanumeric designations are sometimes quirky. I discuss B-26 Marauder testbeds and B-26 Invader testbeds; yes, the Air Force at one time or another called these two different aircraft types by the same nomenclature. Because the B-26 Invader, built by Douglas, served part of its life as the A-26, you will sometimes see Invaders listed as A-26s in these pages. Marauder equals a round cigar-shaped fuselage, built by Martin. Invader equals a angular square-sided fuselage, built by Douglas. And I may occasionally refer to Messerschmitt single-engine fighters as either Me 109 or Bf 109, a nomenclature distinction that will haunt that fighter forever.

Quotations from reports and documentation on these aircraft are employed in this book in an effort to bring the cerebral world of flight test engineers and planners directly to you. Engineer Johnny Armstrong's eloquent description of the X-15A-2 ramjet program is a particular joy to read.

Chapter 1

OLD SCHOOL

Aeronaut Dan Maloney perched in the glider Santa Clara, *ascending with its pioneering balloon mothership in 1905. John J. Montgomery employed hot-air balloons in an effort to give his gliders altitude for extended flights. (The David D. Hatfield Collection/The Museum of Flight)*

Aircraft designers and operators have always faced a battle to achieve greater range or greater altitude to accomplish a mission. In aeronautics the mothership concept to stretch altitude or range is nearly contemporary with the Wright brothers' activities in the early 1900s.

Mother of All Motherships

Pioneer glider designer John J. Montgomery employed balloons to lift his man-carrying gliders to altitude for release as early as 1905 in the vicinity of San Jose, California. Balloonist Frank Hamilton had a hot-air balloon that made these ascents possible.

John J. Montgomery began studying problems of flight in the 1880s in California. Like others of his era, he evaluated birds to learn what he could from their natural structure. He rightly inferred the value of wing camber and came to appreciate what became a traditional airframe design, with aft-mounted tail surfaces deflected to affect pitch and yaw. Montgomery's curiosity and passion for mastering airmanship with a lightweight airframe led him to build gliders starting in an era when engines for such a venture were unheard of. Even after the Wright brothers' seminal powered flight of 17 December 1903, Montgomery continued to experiment with glider airframes.

In addition to his airframe contributions to the state of the art, Montgomery may well stand as the creator of the mothership as a research partner. He had made a number of encouraging glider flights from sloping hills, as had others. But Montgomery was at the forefront of altitude flying, using Hamilton's hot-air balloon to lift a Montgomery glider and aeronaut Dan Maloney as high as 4,000 feet above ground level in an age when other pioneer aviators were barely getting out of ground effect. Although balloons had been used to elevate parachutists before Montgomery's experiments with gliders, Montgomery was on the leading edge of high-altitude flight when he sent his glider and pilot aloft, suspended beneath a hot-air balloon in March 1905. Release of the glider was accomplished by cutting the supporting rope with a knife.

This ingenious altitude assist allowed Maloney to maneuver and turn Montgomery's glider into the wind and in any direction for unpowered flights lasting 15 minutes or more, according to contemporary accounts. Montgomery the inventor needed financial backing for his ongoing flight experiments. Subsequent flights with aeronaut Maloney were part test and part tease as crowds were encouraged, and sometimes charged, to watch.

Tragedy overtook Montgomery's operation on 18 July 1905, when a balloon rope became tangled in the glider, damaging it fatally when aeronaut Maloney cut away from the balloon at about 4,000 feet over Santa Clara. Observers saw the rope catch the glider before release, but their shouting to Maloney to stay connected to the balloon and ride it down as its volume of air slowly cooled went unheard. Maloney may have thought the shouts were merely encouragement for his impending flight.

Viewers noted about 15 seconds after release that the right wing of the crippled glider began to fold. While Maloney struggled to keep the glider under control in its descent, spectators could hear wooden structural members in the airframe cracking. Although he survived initial impact, aeronaut Dan Maloney died about a half hour later. The dangers of test flying manifested themselves chillingly that day over Santa Clara. If motherships were to have utility, safe separation had to be perfected.[1]

Testbed Primeval

Cartoonists have long enjoyed portraying the archetypal caveman inventor chiseling a wheel out of stone, but nobody knows, precisely, who that caveman was. To stretch that delightfully absurd image even further, what about the second version of the stone wheel; was it tested on the caveman's original cart? Long before Internet media made it possible to globally broadcast every event ranging from the most cutting-edge milestone in science to the latest in a ceaseless stream of "Hey, look at this!" moments of mediocrity, it is quite possible that the first testbed airplane, modified from another airplane for the sake of invention, is lost to memory.

The glut of surplus biplanes and engines available in the United States after World War I proved to be a mixed blessing well into the 1920s. The presence of so much surplus might be the enemy of new, costlier aircraft, yet some aeronautical inventors used the tried-and-true airplanes in the quest to further the state of the art. Harlan D. Fowler, along with other pioneer aeronautical engineers, knew as early as World War I that the simple wings and airfoils of that era were compromises in performance that had to embrace landing and takeoff requirements that were significantly different from cruising demands. Fowler's concept of an adaptable wing shape and planform aimed squarely at fixing this limitation by giving the same airplane two significantly different wing layouts for the two major, and different, portions of flight.

By the 1920s, Harlan Fowler was well on his way to perfecting his ideal flap. The Guggenheim School of Aeronautics at New York University had a wind tunnel in which Fowler's design was first tested in model form. By 1927, it was deemed to be ready for actual flight test. Fowler said that his intended testbed aircraft was a Vought VE-7, and his prototype wing was constructed with the load requirements of that airplane in mind. When a VE-7 was not available, Fowler used a small-tail JN-4C Canuck variant of the Curtiss Jenny biplane for his testbed. The resulting wing, although estimated to weigh 100 pounds more than would a wing intended for a Canuck from the outset, nonetheless amply validated Fowler's theory. Even with the heavier wing designed for the VE-7, Fowler's parasol unit weighed 30 pounds less than the original large biplane wings of the Canuck.[2]

Fowler's test pilot was Wesley Smith. The site was the Pitcairn airfield in Philadelphia. Takeoff run consumed 9 seconds in still air; landing to a stop took 10 seconds. Timing the testbed over a measured course, Fowler said that the Canuck with the new wing attained 94 mph, compared with only 74 mph for the original Canuck as a biplane. Ever the perfectionist, Fowler said that this performance boost was attained with no extra effort at streamlining, plus the use of a less-than-optimum propeller on a faulty engine. He opined that rectifying these three things could have given the testbed a top speed of at least 100 mph. Harlan Fowler was excited about the potential of his area-increasing wing flap to translate into greater payloads and/or greater range.[3]

Fowler's flap tested on a modified Canuck was prescient. It took the aircraft industry awhile to catch up, but the Lockheed Model 14 of 1937 and the Consolidated B-24 Liberator of 1939 used Fowler flaps and paved the way for many aircraft that followed.

Piggyback Premier

The rapid amalgamation of aeronautical knowledge and technologies over the decade following John Montgomery's balloon lifts produced viable rigid airships such as Germany's Zeppelins that patrolled the North Sea during World War I. In 1916, the British needed a way to give short-range fighters the ability to reach the enemy Zeppelins, which stayed out of typical fighter range. In what may be the first use of one fixed-wing aircraft to carry and launch another aircraft, a Bristol Scout fighter was mounted to a trimotor Porte seaplane. On 17 May 1916, the aerial combination made a successful flight and separation, but the concept went no further at that time.

The concept of range-extending mothership seaplanes resurfaced in England in 1938. Airmail was the commodity, and long-range aerial delivery that could handily beat an ocean liner was the goal. The mothership this time was a modified Short Empire–class flying boat, designated Model 21, and named Maia. The airmail-carrying aircraft was an unusual four-engine floatplane, the lone Short Model 20 Mercury. Braced with struts, the smaller Mercury rode atop the hefty Maia. The Mercury was otherwise unable to take off with sufficient fuel and payload to make a transatlantic flight, but hitching a ride on the Maia solved the problem.

Following tests in January 1938, the Maia/Mercury combination performed its first separation on 6 February that year. On 21 July, taking full advantage of its airborne perch, Mercury set a world-class nonstop distance record measuring 6,045 miles from Dundee, Ireland, to a spot almost reaching Cape Town, South Africa. Henceforth, the Maia/Mercury lash-up entered scheduled airmail service from Southampton, England, to Alexandria, Egypt, as British aviation exploited the piggyback technology to serve its overseas interests. World War II put an end to the service. Then, extended aircraft range developments made such complex airmail delivery schemes unnecessary in the post-war years.

Parasite combination aircraft form a subset of motherships. (See Chapter 2.) The German Air Force in World War II fielded a number of parasite combinations using a manned single-engine fighter such as a Bf 109 or Fw 190 mounted atop an unmanned twin-engine bomber such as a Ju 88. The bomber featured a radical nose-mounted explosive weapon said to be able to demolish many feet of solid concrete with its force.

The fighter pilot was to fly the combination toward a heavily defended high-priority target. His link to the bomber's controls enabled him to fly both aircraft. Once aim for the target was achieved, the fighter pilot was to lock the bomber's controls, separate from it, and depart the scene with haste.

Total production of these parasite combinations, named Mistel, was around 250. The Luftwaffe massed

The ravages of front-line photography may be to blame for the condition of this photo showing a Mistel combination of Fw 190 and Ju 88 surrounded by rubble and rubbish at war's end. The drastic Mistel program suffered pitfalls, but the concept proved that it could work. (Harry Fisher)

about 60 of them in Denmark for an intended assault on the British fleet, but unsuitable weather foiled that plan. In early 1945, some of the Mistels saw duty against the Soviet advance on Germany. At war's end, both official and unofficial GI photographers captured images of fantastic looking abandoned Mistel combinations at German airfields; their curiosity factor ensured they would be remembered.

Early NACA Testbeds

At Langley, Virginia, researchers from The National Advisory Committee for Aeronautics (NACA) wrestled with problems caused by aircraft icing as they sought ways to mitigate the potentially fatal outcome of ice accretion. Ice could build up on wings and flying surfaces with sufficient weight to make it impossible for an aircraft to climb or even to maintain altitude. Roughness of rime icing degraded airfoil efficiency, and ice could obstruct control movement. Ice accumulation had been known to tear radio masts and antenna wires from aircraft, and ice buildup on propeller blades could create serious out-of-balance vibrations, and the slinging of ice from whirling prop blades could inflict damage on adjacent airframe surfaces. Windscreens obscured by ice might remain opaque through descent to landing. The NACA, charged with finding solutions to aeronautical problems to benefit the growth of aviation in the United States, tackled the ice issue, sometimes employing testbed aircraft.

By 1931, the NACA had mounted a test airfoil shape 2 feet in span and 4 feet in chord to the struts of a Fairchild 71 single-engine monoplane. In this way, icing could be induced on the test airfoil by means of spraying water ahead of it that would not affect the aircraft's actual wings in specific meteorological conditions. This testbed carried a small boiler in the engine exhaust manifold, with a steam line running to the airfoil. The Fairchild sought temperatures as low as 18 degrees Fahrenheit. Water from the spray bar produced icing on the airfoil, which was thwarted by the heat of the steam piped into the test wing section. The tests showed only the forward portion of the wing need be heated to deny ice a foothold.[4]

Validation of the wing-heating premise in the early 1930s with the Fairchild testbed led to the installation of such a system on a Lockheed 12A, followed by the sole XB-24F, an icing research modification of a B-24D Liberator. Continued work at the NACA research center in Ohio in the 1940s resulted in a heavily modified Curtiss C-46 Commando testbed featuring a heated wing and two types of windshield heating. One used warm airflow between two layers of glass; the other employed the radically new NESA-impregnated glass. NESA is an electrically conductive salt that was baked into the windscreen glass. Transparent, NESA can be heated with the electrical resistance of a current wired into the windscreen.[5]

Many of the anti-icing devices tested by the NACA on this C-46 went on to be applied to production aircraft for increased safety.

NACA B-24 Windscreen Icing Testbed

Although B-24 Liberators disappeared from the scene rapidly after the end of World War II, a surprising number were used as testbeds by industry and the NACA both during and after the war. Intensive NACA efforts to combat aircraft icing, plus aircraft engine research, were collocated in Cleveland, Ohio, in this era. One B-24M (44-41986) did double duty while at Cleveland. It served alternately as a testbed for sample heated windscreen panels in a special nose array, as well as carrying a turbojet mounted on a pylon beneath the right wing of the aircraft.

For windscreen icing tests in the frigid winter of 1946–1947, this Liberator flew around the Great Lakes region of the American Midwest with a new nose replacing the standard turret and bombardier's station. Electrically heated windscreen panels surrounded a smaller-diameter bullet-nose shroud and were placed at different angles to the slipstream in an effort to replicate plausible windscreen orientation on aircraft.[6]

Seven windshield panels from Pittsburgh Plate Glass were used in the test. Three of the panels were laminated and electrically heated; the remaining windshield panels were not heated. Two were oriented at a 45-degree angle from the B-24's thrust axis; four were at 30 degrees; and one was at 60 degrees. A special auxiliary power unit (APU) installed in the waist of the Liberator provided electricity to heat the three panels. Transformers allowed the amount of electricity supplied to the heated panels to be varied.[7]

Chapter 2
PARASITES WITH A PURPOSE

The reason for the XF-85's bent tail surfaces becomes clear as the tiny jet is nested in the bomb bay of its B-29 mothership. An add-on landing skid is visible under the nose. A single Westinghouse J-34 turbojet powered the Goblin. (AFFTC/HO)

The term parasite aircraft has come to mean, typically, a mating of more than one flying vehicle for a military purpose. The British experiment with a pursuit mounted atop a giant seaplane in 1916 (see Chapter 1) may be the first true parasite aircraft combination. Great Britain and Germany both experimented with pursuits that could be released from airships during the war, but little came of this before the end of hostilities.

Parasites with a Purpose

After World War I, airship experiments with stowed biplanes achieved a level of repeatable success. The U.S. Navy dirigible *Los Angeles* was the testbed mothership for a trapeze arrangement that proved to be successful at capturing and releasing biplanes in 1929. The following year, *Los Angeles* carried a glider aloft and launched it over Lakehurst, New Jersey.

Curtiss F9C-2 Sparrowhawk (BuNo 9057), flown by U.S. Navy Lt. D. Ward Harrigan, connected to the trapeze mechanism of dirigible USS Macon (ZRS-5) in 1933. (U.S. Navy)

The U.S. Navy dirigible USS Macon conducted initial operations with Curtiss F9C-2 Sparrowhawk biplanes over New Jersey on 7 July 1933. The tiny Sparrowhawks, visible beneath the huge airship, were flown by Lt. D. Ward Harrigan and Lt. (jg) Frederick N. Kivette. (U.S. Navy)

Curtiss XF9C-1 Sparrowhawk (BuNo A8731) is raised into the hangar bay of the dirigible USS Akron (ZRS-4) on 3 May 1932. Pilot is Navy Lt. Howard L. Young. (U.S. Navy/NARA)

By 1931, the first of eight Curtiss F9C biplane fighters was flying. The following year, the F9C went aboard the U.S. Navy dirigible *Akron* as a parasite stowed internally, and lowered for release. The airship *Macon* followed suit as a carrier of F9Cs. F9C pilots quickly learned the mechanics of making launches and recoveries, and the system was more useful than the later XF-85 trapeze attempts beneath a B-29 bomber.

The *Akron* also employed Consolidated N2Y aircraft, two-seat trainers that could use the dirigible's trapeze apparatus. The N2Ys had the ability to serve as couriers and carriers of passengers to and from the dirigible. Size mattered; the N2Y had a wingspan of only 28 feet, and the F9C spanned only 25 and a half feet.

If the airship-and-biplane concept proved to be feasible, the large airships themselves were doomed in a series of crashes that saw both the *Akron* and the *Macon* lost at sea.

Germany's Mistels

There's a fine line, sometimes blurred, between parasite aircraft and mothership combinations. During World War II, the German Luftwaffe sought new ways to get

Chapter 2: Parasites with a Purpose   11

One frame from movie film shows a Sparrowhawk on the Macon's trapeze, left, and a second F9C waiting in a position called the perch, at right, while flying near Naval Air Station Moffett Field, California, on 12 October 1934. Both fighters have their main landing gear removed, as was commonly done for Macon's operations after 19 July 1934. The F9C on the trapeze is flown by Navy Lt. Harold B. Miller; the second F9C is flown by Lt. (jg) Frederick N. Kivette. (U.S. Navy)

combat gliders airborne and delivered to their destination. A series of experiments tested different single-engine airplanes mounted dorsally above a DFS 230 glider. The typical DFS 230 could carry 10 combat troops, including the pilot.

A Klemm Kl 35 monoplane was the first aircraft mounted to the DFS 230. The Klemm could not lift itself and the DFS 230, so a traditional Ju 52 tow plane got the combination aircraft airborne in September 1942. Once released, the test team determined that the Klemm was insufficient to keep the duo airborne, so the two aircraft separated and made safe landings. Other tests with the low-powered Klemm and DFS 230 included releasing the glider at low level, as well as landing with the combination still mated.

Next a Focke-Wulf Fw 56 parasol monoplane with an Argus engine capable of delivering nearly 240 hp was mounted atop the DFS 230. Still insufficient for an independent takeoff, the Fw 56 combination was at least able to maintain altitude once towed aloft.

Tests with the DFS 230 glider continued by mounting a powerful Bf 109E fighter. This necessitated reinforcing the glider fuselage to accommodate the fighter above it, and beefing up the glider landing gear. On 21 June 1943 the Bf 109E–powered combination aircraft took off under its own power from an airfield near Linz, Austria. The parasite combination landed, still attached, after a 40-minute flight. The first attempt at separation later that day, though successful, resulted in the glider pitching up and striking part of the Bf 109 mounting hardware. The addition of an aerodynamic dive brake solved this issue. The Bf 109 rested on a special set of struts, its own landing gear retracted.

Although successful, the pairing of a high-performance fighter with a glider may have been logistically impractical because of the number of fighter aircraft needed for the defense of German targets. (In a similar vein, even though tests of American P-38 fighters as armed glider tugs proved that the concept was feasible, the idea was not adopted at

At the war's end, American troops encountered numerous haunting examples of German technology, including these Mistel Fw 190/Ju 88 combinations at Merseburg, Germany. They were to participate in Operation Eisenhammer, the intended destruction of much of Russia's power generating capacity, but the mission was postponed repeatedly. (USAFA/Brown Collection)

Testbeds, Motherships & Parasites

least in part because of the precious fighter aircraft and combat pilot assets it would divert.)

The German piggyback glider parasite combinations gained the name Mistel, meaning mistletoe, an apt use of a parasitic plant to describe a parasitic aircraft assemblage. The term Mistel continued to describe later German wartime efforts to mount a fighter atop an unmanned twin-engine Ju 88 laden with explosives. The ultimate Mistel bomber featured a heavy shaped charge replacing the normal forward fuselage of the Ju 88. Although potentially devastating if accurate hits could be made, in practice Mistel bombers failed to connect with serious targets in a way that affected the outcome of any combats or campaigns. The Germans took a patriarchal view of their Mistel composites, sometimes referring to them as *Vati und Sohn,* translatable as "Dad and Son." The German word *huckepack,* meaning piggyback (or the British pick-a-back) was also used to describe the Mistel combinations.

A translated German document in the U.S. Air Force Historical Research Agency described some specifics of the Ju 88–based Mistel bomb. Experienced German dive-bomber pilots had used autopilot to aim for an attack, and this worked well in the Ju 88 bomber. "The excellent functioning of this gear in the Ju 88 airplanes produced the idea using this system to place on target these unmanned bomber planes, of which large numbers were held in depots without any prospects of being put to profitable use. For this purpose a manned single-seater Me-109 fighter was mounted above the fuselage of a Ju 88 with releasable struts." To deliver a Mistel flying bomb, the pilot of the Me-109 (or Fw 190 in some cases) flew the composite aircraft to within visual range of the intended target, set the steering on a flat dive from a distance of several kilometers, thereby giving the Ju 88 the proper target heading, course, and glide slope. Still predicated as an anti-shipping weapon, the translated document said: "At a distance of 3,000 to 4,000 yards from the ship, and thus still beyond range of the hazardous multiple AA guns, the fighter pilot disengaged his airplane from the unmanned bomber and changed his own course, while the bomber with its set steering continued on its target course."[8]

Test flights produced deviations from the target of about 11 yards laterally and 22 yards longitudinally, which gave the testers hope that they could expect "good prospects of success in aiming at the silhouette of an aircraft carrier or a battleship," according to the translated document. Because of the high-speed performance of the piloted fighter portion of a Mistel combination, the explosives carried by the unmanned bomber had to be internal (or at least conformal).

The translation explained the development of a hollow charge (shaped charge) explosive for Mistel use: "Efforts had been made to exploit the effectiveness of the hollow charge in composite airplane tactics. The center of gravity was very unfavorably situated in hollow-charge bombs, since they lacked the massive nose section of other bombs and had to be at least of four caliber lengths to obtain stability in trajectory; the SH1-800 was thus already 3.5 meters long. The question now was how long a hollow-charge bomb would have to be to ensure that even at a flat impact angle its blast effect would reach to beneath the armor-plated decks (presuming anti-shipping attacks)."[9]

The ultimate answer to the hollow charge's necessary size and geometry "was found by turning the entire Ju 88 airplane into a bomb: The crew space in the nose was made detachable and replaced by a hollow charge with a diameter of roughly 2 meters and a total weight of approximately 4 tons." The shape of a hollow charge like this includes a hemisphere that helps promote a jet of metal during the charge's explosion that can cut through armor. Copper was considered for lining the hemisphere, but was rejected because it was in short supply in wartime Germany. Instead, hydraulic presses formed sheet-iron hemispheres. "The weight of the finished lining was approximately 1 ton. With an explosive filling of 2.8 tons, and an inner framework and outer casing weighing together approximately 440 pounds, the weight of the entire hollow charge was thus 4 tons."

That weight kept the Ju 88 within center of gravity limitations. The SH1-4000 hollow charge was built with the same diameter as the aircraft fuselage. The high-explosive charge had to be filled as a liquid to ensure good conformance to the lining, with additional powdered ingredient stirred in during the bomb-making process.[10]

After the various Mistel bomb components had been individually tested, the proof came from bombing an actual warship. The German navy had control of an upgraded 1911 French battleship of 25,000 tons, and this was to be the target. Working up to the actual test launch of an unmanned Ju 88 Mistel flying bomb, combinations were made using a manned Ju 88 that could divert and recover after being set on course and released by the Bf 109 pilot. "The crew in the Ju 88 airplane only took over control shortly before reaching the target aimed at, rising to fly over the target, and then landed the airplane," the translated document explained. "Once the composite was properly adjusted, the crew cabin was removed from the Ju 88 . . . and replaced by its warhead, in this case the SH1-4000 hollow charge." In this way, it was planned that the Ju 88s could be adjusted and even ferried, with the hollow charges stored at tactical airfields where they would be needed for operational missions.[11]

Before the target French battleship test, a number of live Mistel combination aircraft took off from the German research center at Peenemünde and headed for the Danish island of Moen. At Moen, a steep cliff about 300 feet high was fitted with a triangular target at its mid-level. The German source document says, "In these exercises all airplanes released scored hits within 20 meters from the center of the target triangle, which amounted to 100-percent hits in an attack launched diagonally at the longitudinal axis of a large ship." The Number 66 electric impact fuses in the noses of the hollow charges worked well in the tests. Testers noted that the Ju 88 autopilot kept the flying bomb "firmly on course." German photographers at sea captured the impacts, as did a special Fieseler Storch flying overhead with a vertically mounted camera.[12]

Impetus for anti-battleship Mistel development at this time in early 1944 was provided at least in part by German planning for the anticipated Allied invasion of France. In March 1944, final tests were made with the expendable French battleship. A shaped charge of the SH1-400 type was placed against a forward gun turret to test its capability on a ship. Results were promising as the charge penetrated the two turrets and exploded shells that had been placed inside. With the French battleship anchored near Toulon, Mistel aircraft launched from Marseilles in a realistic combat run. The unmanned Ju 88 struck the ship's stern at an angle of less than 20 degrees to its centerline, and about 20 degrees off horizontal. A huge hole opened up in the ship's side, blasting out an armored steel bulkhead 270 mm thick.

Germany's Mistel concept for piggyback aircraft began with a quest for an alternate means of delivering combat gliders to the front. By the war's end, Mistel had produced a fighter-and-bomber lash-up, using a piloted Fw 190 as in this snapshot, or alternately a Bf 109, mated with an unmanned Ju 88 flying bomb. This abandoned Mistel combination was photographed by advancing American troop carrier crews. Even for an intended one-way desperate sortie, someone took the time to apply irregular camouflage to the struts linking the fighter and the bomber. This Mistel combination is probably part of the equipment found at Merseburg that was intended for the German Operation Eisenhammer. (Joseph Obendorf Collection)

"Presumably," the German document says, "the shot went through the bottom of the ship, which sank in the shallow water."[13]

Success looked complete for the shaped-charge Mistel. But now, the German air force general staff was less interested in using Mistel to repel the invasion because estimates indicated much smaller ships and landing craft would comprise the Allied invasion fleet. Although the actual D-Day landings made extensive use of smaller landing craft, big battleships did send large rounds miles inland, to the consternation of the Germans. One document says that this battleship shelling caused an order to go out, too late to be of use, to ready 300 SH1-4000 hollow charges for sorties against them.

Germany also embraced the Mistel arrangement as a means of putting a rocket-powered reconnaissance aircraft at altitude high enough to permit the rocket engine to further elevate the reconnaissance vehicle above enemy interception. The DFS 228 was designed with a pressurized nose capsule for the lone prone pilot, a center section for infrared

14   Testbeds, Motherships & Parasites

cameras, and a tailcone accommodating the nozzle of the Walter bi-fuel rocket engine.

The DFS 228 only attained glider flights before war's end, but it validated the use of a Dornier Do 217 twin-engine bomber as a manned mothership to get the DFS airborne. The ambitious mission profile envisioned for the powered DFS 228 called for the Do 217 to carry the dorsally mounted reconnaissance aircraft to 33,000 feet where it would be launched, the rocket then being fired to give the DFS 228 its design altitude of 75,000 feet. Reconnaissance photography was made at extreme high altitude with the DFS 228 in a shallow glide. Between the rocket-powered and glider portions of its flight, the DFS 228 was expected to have a range of 650 miles. Already on the drawing board at DFS when the war ended was a supersonic follow-on, the unbuilt DFS 346.

After World War II, the projected global range of strategic bombers such as the giant B-36 posed potential problems for fighter escorts. The U.S. Air Force studied several possible efficiencies to tackle this, involving fighters hitching a ride on a bomber.

Betty Takes Baka to Battle

In the accepted Allied code naming for Japanese aircraft in World War II, Betty referred to the Mitsubishi G4M series twin-engine bomber workhorse. Baka, another Allied code name, was Japanese for "fool," and was the moniker assigned to the Yokosuka MXY7 Ohka (Cherry Blossom) piloted bomb. Betty and Baka collaborated in a desperate effort to cripple U.S. ships during the assault on Okinawa.

Japanese military tradition and culture acknowledged the use of suicide attacks as potentially valuable tools of war. This made for a fairly easy segue into building a piloted suicide flying bomb to defend Japan. In the summer of 1944, as Allied forces drew ever nearer to Japan, Ensign Mitsuo Ohta drafted a notional rocket-powered suicide aircraft. By August of that year, Ohta submitted his proposal, which gained favor with the Japanese Navy. Detailed plans and engineering work followed quickly under the designation MXY7. Initial thought was to use the MXY7 as a coastal defense weapon, an anti-invasion tool. A mothership was required to get the MXY7 aloft and transport it to within gliding, diving, and limited rocket power range of its intended target.[14]

Everything about the Ohka project was simple and streamlined. Construction was largely of wood and intended to be readily made by unskilled workers. Cockpit instrumentation was limited; a one-way flight to the target demanded nothing more. As of the end of September 1944, 10 of the flying bombs were already built. The Model 11 Ohka was fitted with an explosive warhead of 1,200 kg (more than a ton). One Ohka could be shackled in the bomb bay of a Mitsubishi G4M2e version of the Betty, with bomb doors removed. Three rocket engines in the tail could be ignited singly or simultaneously to provide a combined thrust said to be 1,764 pounds for a period of about 8 to 10 seconds.

Glide flight tests began at Sagami in October 1944. In November, the first powered Ohka flight took place at Kashima. Some tests were unmanned. With all three rocket engines firing, Ohka could exceed 400 mph. Through March

Japanese pilots photographed near a Betty bomber carrying an Ohka piloted bomb in 1945. (Naval History and Heritage Command Photo NH 73100)

Chapter 2: Parasites with a Purpose   15

An American military guard protects an Ohka Model 11 piloted suicide bomb at Yontan Airfield, Okinawa, in 1945. Mitsubishi Betty bombers were capable of serving as motherships for the Ohka. (U.S. Navy/NARA)

1945, 755 Ohka Model 11 Navy Suicide Attacker flying bombs were constructed.[15]

On 21 March 1945, less than two weeks before the start of the Battle for Okinawa, the 721st Kokutai (Naval Air Group) sent 16 G4M motherships each carrying an Ohka into battle, but interception led to the premature release of the hapless pilots and their Ohka bombs. On 1 April 1945, Ohkas damaged the battleship USS *West Virginia* and three transport ships. On 12 April, an Ohka aimed at the American destroyer radar picket ship USS *Mannert L. Abele* off the shore of Okinawa. The Ohka pilot found his mark, the flying bomb exploding with such violence that the destroyer, already damaged by another suicide attack, was cleaved in two. Ohkas later hit two other picket ships, but only the destroyer *Abele* succumbed.

The weight of the Model 11 Ohka made the G4M motherships slow and cumbersome. New piloted bomb designs attempted to rectify that by carrying smaller explosives. A faster mothership, a modified Yokosuka P1Y3 Ginga twin-engine bomber, was in design to carry the later Ohkas, but it remained a paper airplane by war's end. An earlier Ginga prototype became a testbed for the Tsu-11 Campini-style jet engine designed for the Ohka Model 22. The Model 22 was too late for combat.[16]

A turbojet, the Ne-20, was to have powered the Ohka Model 33. The Ohka 43A, also to be Ne-20 jet powered, had folding wings for stowage on deck hangars of special Japanese submarines. The Ohka 43B was to be secreted in caves and launched by catapults in defense of the Japanese home islands. None of these later variants saw use. The intent was for the unfulfilled Model 33 Ohka to be carried and released by the G8N1 Renzan attack bomber, but that bomber's low developmental priority led to cancellation of the Ohka 33.[17]

The U.S. Navy kept statistics on the Japanese aircraft challenging Navy operations. During the Philippines offensive, the Navy counted 24 percent of Japanese attackers as suicide attacks. In the drive for Okinawa, the number of Japanese suicide attacks rose to 31 percent of all Japanese aircraft encountered by U.S. Navy ships. The Navy counseled the fleet in 1945 to expect the rate of suicide attackers to continue to rise. It was expected that Japan would husband its remaining aircraft to use specifically in suicide attacks against troop delivery ships in an effort to hurt the anticipated invasion of Japan. Part of the mix of the actual and anticipated suicide attacks was Ohka-Betty combination aircraft.[18]

The Navy had a secret weapon that had already proved its worth (the proximity fuze or Variable Time [VT] fuze) that was compatible with 5-inch anti-aircraft guns on many Navy ships. By reaching out to the range of the 5-inch guns with VT-fuzed rounds, all Japanese attackers faced a serious curtain of accurate fire. The records showed suicide attackers had a greater percentage of contact with their ship targets than did conventional Japanese attackers, and the Navy calculated this would encourage more use of suicide attacks, be it Ohka carried bombs or regular aircraft on a one-way sortie. And the use of Ohka (or Baka in the Allied naming system) manned flying bombs was considered an improvement in Japanese suicide attack capabilities, according to a U.S. Navy study at the time.[19]

Ohka operations were of increasing concern to the U.S. Navy in the summer of 1945. Not used extensively in the Okinawa campaign, nonetheless 10 Ohka attacks were made on surface vessels in April and May 1945, resulting in three hits plus a damaging near miss and six non-damaging misses.

A Navy assessment said, "The Baka . . . presents the most difficult target our surface forces and aircraft have encountered in the war to date. It is also potentially the most dangerous anti-shipping weapon to be devised, being a guided missile with the best possible control, a human being."

A prisoner of war told the Navy that the ideal release altitude from the Betty bomber was 23,000 feet. The Navy calculated impact speeds for the Ohka to range from 525 to 618 mph, based on either a horizontal approach or a 45-degree dive angle. (Another credible post-war source lists the Ohka Model 11's terminal dive speed as 576 mph.) These speeds made the Ohka difficult for shipboard anti-aircraft batteries to track. The Navy calculated, "At a gliding speed of 300 mph and rocket speed of 550 mph, an approaching Baka would be within firing range of various weapons for periods as follows:

- 5-inch/38: 53 seconds (10,000 yards in)
- 40-mm: 13 seconds (3,500 yards in)
- 20-mm: 5.5 seconds (1,500 yards in)"[20]

Ignition of the Baka's rocket engines added acceleration into the firing problem. High speed and small size made the Ohka/Baka difficult to hit: "Although the Baka is unable to maneuver radically, its speed is so great that a slight change in course will create errors as large as if a slower target were maneuvering radically." The good news for the Navy was the fact that the small Baka, with plywood wings, would still trigger a VT fuzed anti-aircraft round at a radius of from 30 to 45 feet away. "Because of its high speed, however, approximately four times as many bursts as against a conventional aircraft may be required to register a kill." Part of the difficulty in downing an Ohka was its lack of an "engine, propeller, or gas tanks, all vulnerable parts of a standard aircraft."[21]

The plan for countering Ohka flying bomb attacks was to intercept the bombers carrying them as soon as possible, with friendly fighters downing the bombers or possibly causing them to release the flying bombs too far from their intended targets. Fighters from Navy Task Force 58 downed 32 Betty motherships carrying Ohkas on 21 March 1945. In an environment where increased Ohka attacks were expected, the Navy advised fighters to give priority to targeting inbound twin-engine aircraft. When in glide mode, the Ohka was within approach speeds by fighters, so attacks

Although the imagery from 16-mm gun cameras leaves something to be desired, this is a rare reference from the U.S. Navy. These photos depict attacks on G4M2 bombers carrying Ohka piloted bombs on 21 March 1945. (U.S. Navy)

on released Ohkas at long range could succeed. At the least, these encounters might cause the Ohka pilot to use the acceleration rocket engines early, possibly depleting the speed advantage before arriving at the target ship. The Navy rightly predicted that a dearth of motherships could lead to Ohkas being ground-launched where feasible.[22] However, the climactic end of World War II in August 1945 pre-empted an ultimate Ohka showdown at Japan's doorstep.

As an aside, the almost anecdotal, offhand references to using a Ginga bomber as a powerplant testbed show Japan to be in the global vanguard of nations using multi-engine aircraft for this purpose.

Chapter 2: Parasites with a Purpose

A Seattle-built B-17G (43-39119) carried a pair of JB-2 buzz bomb knock-offs during USAAF tests at Wendover, Utah. (Glen Edwards Collection via AFFTC/HO)

V-1 and JB-2 Hitch a Ride

In the winter of 1943 verging into 1944, Germany tested the Heinkel He 111H twin-engine bomber as a mothership for a single Fi 103 (V-1) buzz bomb, flying out of the vaunted Peenemünde rocket test complex. If fixed-rail V-1 launch sites in France were vulnerable and elaborate to construct, an alternative was to air launch a V-1 from a Luftwaffe bomber. Pragmatic Luftwaffe planners had to acknowledge the potential of losing the fixed V-1 sites in France to Allied bombardment and even invasion. The venerable Heinkel He 111 was updated through iterative models to become the He 111H-22 carrier aircraft. Curiously, the offset mounting for the single V-1 is seen under the left wing inboard of the engine on a test He 111, and under the right wing on an operational Heinkel from KG 3. At Oschatz, Germany, a modification center was established to convert some He 111H-16s and He 111H-20s into V-1 carrier aircraft.[23]

Missions were flown under cover of darkness at an altitude of 1,500 feet, aiming the Heinkel and its buzz bomb toward a city, where the chances of the V-1 plunging to earth and exploding with some effect was considered worth the effort. Part of Kampfgeschwader 3 went operational with the Fi 103 sorties from Holland in late July 1944, aiming the flying weapons toward London and Southampton. The selection of He 111s as the carrier aircraft was probably intentional. By that stage in the war, He 111s were performing an increasing number of non-bombing missions for the Luftwaffe, and it was probably easier to get He 111s released for carrier aircraft conversion.[24]

A contingent of AAF B-17Gs was modified to carry a JB-2 American copy of the German buzz bomb under each wing, with tests made over Eglin Field ranges in Florida as well as the vast American West, flying out of Wendover Army Airfield near the Nevada-Utah boundary. Fresh from combat in A-20 Havocs, a young Captain Glen Edwards participated in the Wendover JB-2 tests. He wrote his impressions of the operations, "We started the jet going and launched the bombs. They dropped away and as they picked up speed, they leveled out and went on their merry way. Jet bombs such as these were guided or flown by an automatic pilot. The pilot was set to nose them into the ground after a certain amount of time was gone, or the fuel was used up. . . . Sure was hot at Wendover."[25]

Just a few years later, Glen Edwards died in the crash of the Northrop YB-49 Flying Wing. Edwards Air Force Base is named in his honor.

In 1947, the Air Force's First Experimental Guided Missiles Group at Eglin produced a set of recommended procedures for crews operating B-17s carrying JB-2s. In addition to the B-17's aircrew, a missile crew tended to the JB-2s. One crewman had responsibility for manual release of the Loons, and he and one other JB-2 team member were poised for this on takeoff in the event of emergency action required by the bomber pilot.[26]

The procedure said that normal release of the JB-2s started with the missile under the left wing of the B-17. When it was released, the mothership was to make a 180-degree turn to the left, or empty, wing. Another 180-degree turn into the same empty wing would bring the mothership back to its intended heading, preparatory to releasing the second JB-2. For the test drops made over the ocean from Eglin Field, an armed fighter picked up the chase of the launched JB-2, escorting it and relaying flight data to the launch

The rigors of spartan desert life at Wendover Army Airfield near the Nevada-Utah border were experienced by the Army Air Forces (AAF) team testing JB-2 buzz bombs released from B-17G aircraft, including the one behind the men in this group photo from the album of Capt. Glen Edwards, who later became the namesake for Edwards Air Force Base. Captain Edwards is third from the right in the back row, wearing sunglasses. (Glen Edwards photo album via AFFTC/HO)

From Glen Edwards' photo album, he said this about the JB-2 project: "The jet bombs were the same as the German V-1 bombs that were rained on England before the end of the war. My project was to determine how much two of them slung under the wings affected the performance of the B-17. They weighed 5,000 pounds each and were quite a strain on the wings. The results indicated they slowed the airplane down 20 mph. The object of the jet bomb project was to fly to within about 100 miles of an enemy target, launch the bombs and go home. We never used them. [The] war ended."

Chapter 2: Parasites with a Purpose

With cartoonish artwork, this JB-2 waits for its one-way mission out of Wendover. The full-up weight of a JB-2 hanging from the wing of a B-17 was 5,000 pounds. (Glen Edwards Collection via AFFTC/HO)

aircraft. The JB-2's mission was intended to terminate in one of two ways, either a predesignated splashdown in the ocean or a shoot-down by the escorting fighter (sometimes a P-80 Shooting Star).[27]

For proper JB-2 operations, it was necessary to ensure a drop speed of 200 mph, sometimes causing the B-17 mothership to enter a dive before releasing a JB-2. Although B-17 mothership operations with JB-2s were fairly short-lived in the Air Force, at the time of the October 1947 Eglin air launches, testers recommended painting the upper surfaces of the silver B-17G wings to improve visibility of the JB-2s by scanners, whose vision could be degraded by the dazzling reflected sunlight off the natural aluminum surfaces of the B-17 wings.[28]

XF-85

The most ambitious venture in terms of new hardware was the XF-85 Goblin parasite fighter. The Goblin's inception dates back to wartime 1944. The Army Air Forces (AAF) was considering the eventual availability of super bombers like the B-36, then in design. Jet fighter escorts envisioned for that era would lack range sufficient to escort the B-36, so the AAF asked the industry to propose parasite fighters that could be stowed aboard the big bombers of the B-35 and B-36 class.

The young McDonnell Company initially responded with a fighter design that was semi-concealed in the bomb bay of a super bomber. The AAF wanted full internal stowage, so McDonnell engineers crafted a smaller version of their parasite fighter, with a wingspan barely more than 21 feet, swept back at 34 degrees. The wings folded to fit inside the bomb bay. A geometric puzzle of a tail performed its aerodynamic duties while being small enough to fit inside the mothership. Perhaps most telling about the constraints the designers faced is the Goblin's overall length: only 14.8 feet.

The parasite had a pressurized cockpit with an "ejector seat," in the parlance of the day, and self-sealing fuel tanks. In the nose, ahead of the windscreen, a large retractable hook provided the means for attaching to the mothership's trapeze apparatus.

The Goblin was powered by a single Westinghouse J-34 jet engine expected by McDonnell to give the egg-shaped fighter speed well over 500 knots; this may have been an optimistic notion.

The Army Air Forces funded two prototypes of the XP-85 (later XF-85) Goblin. A test B-29 Superfortress (44-84111) was modified with a trapeze mechanism to stow and release the Goblin in flight. Photos of the special B-29 show the nickname *Monstro* painted on the bomber, a reference to the menacing whale that swallowed Pinocchio. So constrained was the tiny Goblin airframe that no provision for landing gear was made. The parasite fighter's life consisted of a ride outbound in the bomb bay, aerial combat as needed, and a ride back in the bomb bay, landing while mounted inside the bomber host.

At Muroc, XF-85 assembly was completed in June 1948 following shipment of the aircraft. Ground engine runs and trapeze modifications were also made at Muroc that month. In July, the B-29 mothership made a flight for a function check of the trapeze, which was extended and retracted with the XF-85 attached.

The combat rationale said that the Goblin would be able to return to its mothership to refuel and rearm twice during a mission. Five tons of equipment, fuel, and supplies were to be stowed in the B-36 for Goblin support. A declassified description of the XP-85 spelled out requirements for the pilot: "The physical limitations on a pilot are 5 feet 8 inches and 200-pound weight with full equipment."[29]

The time required to launch and retrieve a Goblin was estimated at 2½ minutes. But this computation was made before any test flights took place; the reality proved to be something different. While the XP-85 was still in gestation at St. Louis, the AN/APN-61 radar beacon was under development to enable the XP-85 to home in on its mothership.

The conceptual rationale included the basic plan that one Goblin could be stowed in the number-1 bomb bay of a B-36, leaving additional bay space available for a smaller bomb load. Conceivably, some B-36s in a group effort could carry up to three Goblins instead of bombs. What was left unspoken is the fate of a flying Goblin whose bomber was downed while the parasite was away.

The two prototype XF-85s had a feature not intended for production versions: skids to allow emergency landings if testing went awry. And it did. On 23 August 1948, during the first flight of a free-range Goblin, pilot Ed Schoch barely avoided serious injury when part of the trapeze smashed the Plexiglas canopy as he tried to dock with the B-29. The skid enabled Schoch to recover safely on the dry lakebed below.

An 8 March 1949 flight damaged the trapeze; the Goblin again made a safe skid landing on Rogers Dry Lake. Skid landings in the F-85 could chew up 1,400 feet of dry lakebed, leaving a trail of fine dislodged clay dust. Fewer than half of the handful of Goblin free flights resulted in hookups and landings with the B-29 *Monstro*.

The final Goblin flight was made on 8 April 1949. This was actually the first free flight of the number-1 Goblin. After three attempts to engage the trapeze, the pilot made a lakebed landing. The Muroc Air Force Base historical records for that period say that the Goblin program "was put on inactive list pending modification of trapeze."

The Air Force pondered the Goblin test experience. If test pilots were selected because they were better than average, and if they had major difficulties docking with the bomber, how would rank-and-file operational pilots make out? The performance of the stubby Goblin would likely fall behind that of conventional foreign jet fighters soon to be fielded abroad. For these and fiscal reasons, the XF-85 Goblin parasite fighter program was ended by the Air Force in October 1949. Some contemporary Air Force documentation referred to the XF-85 as the Internally Stowed Fighter.

Before it ever flew, on 15 July 1948, the XF-85 Goblin posed on the ramp at Muroc for this portrait with a helmeted pilot in the cockpit. Monstro, the XF-85's B-29 mothership, is visible behind the Goblin. A dolly was the only means to move the Goblin on the ground; it had no landing gear. (AFFTC/HO)

An XF-85 is attaching to a B-29 and hoisting into its bomb bay position. The date, 15 July 1948, shows that this operation is taking place in the month before the first Goblin release flight. Tests, including a July flight in which the Goblin was simply lowered and raised while attached to the trapeze, preceded actual inflight release. (AFFTC/HO)

This photo, taken in the XF-85 concrete loading pit at Muroc, shows the Goblin nested in the B-29's bomb bay on 2 August 1948. (AFFTC/HO)

XF-85 in free flight over California. Docking could be tricky; a number of test sorties ended in landings on the dry lake bed when reconnecting with the B-29 proved to be problematic. (AFFTC/HO)

A straight crease in the reticulated surface of the dry lakebed defined the end of Flight 524-6 for the XF-85 when a hookup with the B-29 mothership could not be effected. (AFFTC/HO)

22   Testbeds, Motherships & Parasites

The ovoid XF-85 featured a large hook located ahead of the pilot for engaging the trapeze mechanism on the experimental B-29 mothership. (AFFTC/HO)

In 1948, the XF-85 and its B-29 carrier aircraft were photographed in front of the special pit that enabled the B-29 to load and unload the wheelless fighter. In this era, pre–Edwards Air Force Base, Muroc looked more like the B-24 training base it had been during much of World War II. Most of the wooden buildings in this photo have long since been demolished and this part of the field, now called South Base, hosts modern facilities for testing the latest aircraft and systems. (AFFTC/HO)

FICON

The next time the Air Force wanted a fighter in the bomb bay of a B-36, the system compromised to allow a mostly intact F-84 or RF-84F to ride partially inside, with its wings and especially downward-canted elevators outside.

Dubbed FICON for fighter conveyor, the system attained limited operational use in the 1950s.

With the capabilities of a full-up jet fighter ensured, the next order of business was to design a trapeze mechanism for the B-36 that could extend far enough below the fuselage to minimize the chance of unintended

Early FICON testing used a modified YF-84F (49-2430) coupling with this JRB-36F (49-2707). Before FICON achieved limited operational capability, the trapeze was altered to improve its usability; the aircraft changed to a reconnaissance model of the F-84. (Convair via AFFTC/HO)

Chapter 2: Parasites with a Purpose   23

There's more than one way to mount a FICON. Convair made a set of ramps to raise the mainwheels and aft fuselage of a GRB-36 while leaving the nose gear on the ground. Alternately, at Fairchild Air Force Base near Spokane, Washington, a pit was cut into the ramp to place the RF-84K low enough for the B-36 to be towed over it. Because the F-84 part of FICON had fully operable landing gear, unlike the XF-85 program, FICON crews sometimes opted to take off separately and unite in flight. (AFFTC/HO)

The mated FICON combination could take off, but over-rotation of the B-36 was to be avoided because portions of the RF-84K were in close proximity to the pavement. This image shows a mated FICON setup at Edwards Air Force Base on 23 April 1956. (AFFTC/HO)

contact. The first attempt at hooking up early in 1952, however, encountered instability and porpoising with the F-84. When it became impossible to disengage from the trapeze for some reason, a set of emergency explosive bolts saved the day and allowed the F-84 to exit. A redesign of the trapeze mechanism ensued in an effort to make it more rigid. An outcome was the shortening of some portions, and relocation of the F-84 hook, that put the fighter pilot in closer proximity to the action. But when first tested, an aft locking pin on the trapeze device sheared, necessitating another emergency escape for the F-84 and its pilot.

Several pilots attested to the difficulty of the docking maneuver, so Convair again redesigned the system. Now, docking a modified F-84 with a B-36 was deemed to be feasible.

In the hectic Cold War pace of military aviation advancements on both sides, the time period between the first Air Force experiments with parasite fighters and the validation of the FICON concept saw the practicality of carrying a fighter for bomber defense diminish. But a new arena, high-altitude reconnaissance, had needs and requirements in the 1950s that might be well-served by such a lash-up. It seemed as if the RF-84 Thunderflash photo reconnaissance jet was perfect for such a scheme. Small and fast, the RF-84F (redesignated RF-84K for FICON) could dash into hostile airspace and back out again, hitching a ride both ways in the belly of a giant B-36.

An intriguing bonus came with the RF-84K; it was already certified as nuclear capable. Conceivably, an RF-84K also could be used as the final sprinter in a relay delivery of a nuclear device to a target, again hitching a ride on a B-36.

The FICON program also generated a sunken loading/unloading pit at Fairchild AFB, Washington, where the 99th Strategic Reconnaissance Wing operated the GRB-36D motherships. Although the RF-84K could be mated and demated in this way on the ground, the clearance of a loaded GRB-36

The Q-14 coupled to the C-47 for towing tests circa 1950. This combination paved the way for tests with Republic F-84s and a B-29. (NARA)

was so tight that in-flight hookups and separations were preferred.

By late 1955, Air Force FICON hookups were under way. Never easy, they were, nonetheless, considered operationally viable. The end of the FICON program coincided with the winding down of B-36 operations in the last half of the 1950s and the introduction of the specialized U-2 reconnaissance aircraft. By 1956, FICON use was discontinued, and some of the droop-tailed RF-84Ks finished their military careers as traditional RF-84s, using runways instead of trapezes.

Tip Tow and Tom-Tom

Two other U.S. Air Force fighter hitchhike mothership or parasite program ideas deserve mention. A range-extending notion was the attachment of a smaller aircraft to the wingtip of a larger one, letting the smaller one hitch a ride. The genesis of this was German wartime theoretical

Chapter 2: Parasites with a Purpose   25

The Q-14 coupled to the wingtip of its C-47 mothership circa 1950. Wingtip coupling promised aerodynamic efficiencies produced by the combined aspect ratio of the wings of the joined aircraft. (NARA)

development that proposed adding wing panels to an aircraft to increase its span as an efficiency measure. By increasing span, the wing's aspect ratio is favorably changed, which causes a decrease in induced drag. In the immediate post-war U.S. Air Force, efforts to give jet fighters sufficient range to escort global bombers saw several ideas pursued. Two German expatriates at Wright Field, Dr. Richard Vogt and Ben Hohmann, helped promulgate a two-tiered test of the wing-extension concept.

Beginning in 1949 in the United States, a C-47 wingtip-to-wingtip coupling with a small Culver PQ-14B Cadet showed that it could be done. The C-47 and Culver Cadet were selected for this proof of concept demonstration because their relative sizes had the same ratios of mass and wingspan as would a full-scale demonstration project using straight-wing F-84 Thunderjets attached to the wingtips of a B-29 Superfortress.

The hookup mechanism for this early, scaled prototype was a master of simplicity. The right wingtip of the C-47 received some localized reinforcement to hold a circular ring on a small extension boom, with the ring's opening positioned vertically to the line of flight. The PQ-14B (sometimes referred to as simply Q-14) mounted a rearward-facing lance on its left wingtip. The plucky Culver pilot inserted the lance into the O-ring, letting drag keep it inserted; no securing latch was provided in this early test configuration. Adding power to the Q-14 enabled it to accelerate out of the ring. The C-47 also had the ability to instantly jettison the wingtip ring in an emergency.

Although simplicity was preserved with this arrangement, it placed additional demands on the Culver pilot, who had to back his aircraft's lance into the O-ring on the C-47 wingtip. So simple and austere was the initial C-47/PQ-14 hookup that there was no funding for wind tunnel tests as predictive tools. The test team speculated that the Q-14 might be able to remain stable while connected to the C-47's wingtip mainly by the Culver pilot's judicious use of aileron control. Others wondered if the Q-14's elevators would come into play.[30]

The pilot of the Q-14 was Maj. Clarence E. "Bud" Anderson, a wartime P-51 triple ace. The first coupling effort was made on 19 August 1949. Years later, Anderson described that first attempt at coupling in flight: "Preliminary flights in close proximity to the C47 tip with the coupling mechanism installed revealed the strong vortex airflow about the wingtip, which made precise close formation flight quite difficult. For the initial try, the C-47 climbed to an altitude of 8,000 feet for safety considerations. At that altitude the Q-14 had little or no excess power for maneuvering in and out of position, and this proved to be a problem."

During this first hookup attempt, the receiving ring on the C-47 was positioned about 3 inches from the transport's wingtip. This proved to be problematic due to the C-47's wingtip vortex flow. Anderson finally made a successful connection in the turbulent air at 8,000 feet by cutting back the Culver's throttle more quickly than he had wanted. The Q-14 engaged the C-47 ring, and immediately the much larger aircraft's wingtip vortex caused the attached Culver to roll right, dropping to a 90-degree orientation to the C-47, which remained in level flight. Anderson said, "This didn't seem right, so power was advanced for immediate disengagement. The whole thing happened in a few seconds. No further attempts were made that day since a slight bend was observed in the C-47 wingtip."[31]

After this event, the C-47 ring was moved out 19 inches from the wingtip on a longer boom. The next coupling flight took place on 7 October 1949. Test pilot Anderson made contact and only partially reduced power on the Q-14, and interestingly, the elevators helped with roll control while the Q-14 was connected to the mothership C-47. "At first it was difficult to adjust to the use of the elevator for roll control since the normal reaction was to use ailerons for roll about the coupled point," he said. "Four separate couplings were made that date and the longest coupled flight was 5 minutes in duration."

As Anderson gained experience and confidence with the hookup procedure, the effort became almost routine. One help was to conduct the coupling at the lowest altitude where calm winds could be found, giving the Q-14 better performance. Anderson described the procedure:

"The C-47 would stabilize at 95 knots, which gave the Q-14 adequate reserve power to maneuver in and out of position. Couplings became very easy, and with the increased experience, average time required to hook up was somewhere between 10 and 30 seconds. Once hooked up, power was reduced to idle and control of the Q-14 was maintained by the elevator exclusively. The ailerons were almost completely ineffective, and a very slight elevator movement could easily override full aileron. Once coupled, the C-47 could increase airspeed to a normal cruise of around 120 knots, where coupled stability actually improved. At that point the C-47 could do about anything that was normal: roll, climb, and dive in smooth air, turbulence, or weather. Even night couplings were made as part of the test program."[32]

Once Anderson had mastered the ability to connect his diminutive Q-14 to the C-47's wingtip, some couplings at night broadened the envelope even more. However, some things remained beyond the aircraft's capabilities under the circumstances. "In the Q-14, hands-off flying was never possible for more than a few seconds," Anderson explained.

Further test iterations included disabling the Culver's inboard aileron, the aileron most affected by the C-47's wingtip vortex, but control was still considered inadequate. Anderson described further tests, made possible by the increased confidence in this simple wingtip hookup system: "To prove a point, the engagement ring on the C-47 wingtip was moved back to the original 3-inch position, and successful couplings were demonstrated under the influence of the stronger wingtip vortex.

"One point of interest was that once the Q-14 pilot gained full confidence, he could maintain position by flying instruments, if he so desired. All you had to do was to watch the turn/bank indicator and keep the ball centered with the elevator."[33]

Into 1950, the C-47/Q-14 combination logged 231 couplings and 28.5 hours of towed flight time. "A large sampling of pilots was taken also; mostly fighter pilots, but some multiengine pilots were used also. Only one pilot failed to make a midair coupling," Anderson said.

The longest coupled flight was 4:08 hours. These early hook ups gave the testers confidence to continue with the full-scale B-29 and F-84 coupling operation, but many questions remained.

The rudimentary boilerplate nature of the C-47/Q-14 links meant that the flights were made without the ability to measure or quantify performance data.[34] While the C-47-and-Culver experiment evolved, a contract was already made with Republic Aviation to modify a B-29 Superfortress with much more complex wingtip mounts for a pair of F-84D jet fighters.

Tip Tow Project

Project MX-106 was the B-29/F-84 test effort, also known as Tip Tow. The B-29's outer wing panels were replaced with sections containing the wingtip towing and retrieving mechanism. The connections were much more sophisticated than the simple ring-and-lance system previously used. The B-29 mount had provision to allow some movement by the F-84 in all three axes, pitch, roll, and yaw. Rubber pads on each B-29 wingtip were intended to provide a seal between bomber and fighter wings, conducive to greater inflight efficiency. The two F-84s were modified to accommodate the forward-facing lance on a wingtip. An effort was made to have a hydromechanical connection between the lance and the F-84's aileron controls. Anderson explained, "It was hoped that this simple method, which was based on a ratio of lance rotation or flap angle to aileron movement, would provide automatic control."

The F-84 pilot could quickly override the system by manual control inputs or by rapidly disconnecting the aircraft. Once the B-29 and F-84 were fully mated, locks limited the fighter's axes of movement. Now the F-84 only had separate movement from the B-29 in the roll axis, or "flapping," as the F-84 could move in a rolling arc while connected to the B-29. Explosive bolts separated the two aircraft automatically if flapping limits were exceeded. Or, the bolts could be detonated intentionally from either aircraft if needed.

For Tip Tow, only the right-hand F-84 was instrumented to quantify valuable flight test data; the left-hand F-84 had all of the hookup equipment, but not the data recording gear.[35]

The Tip Tow method of connecting two F-84 Thunderjet fighters to the wingtips of a B-29 was a way to increase the range of early jet fighters. A catastrophic collision and fatal crash with this system was devastating to the program. (AFFTC/HO)

The first link-up in flight between an F-84 and the B-29 was on 21 July 1950. Successful engagements ensued. Testers soon learned that there were differences between an F-84 jet fighter and a Q-14 manned drone. Although the F-84 had more power, early turbojets were notoriously slow to spool up and transfer their power into useful thrust. "A smooth, steady lineup and insertion of the lance produced satisfactory hookups," Anderson said.

The potential for oscillation of the B-29's more flexible wing was a possibility that caused the testers to avoid rough air whenever feasible. This also argued for moving the connected F-84 into the locked-down position expeditiously, because until this happened, the fighter could move about all three axes, possibly introducing dangerous structural oscillation in the B-29 wing. Once the F-84 was in the locked position on the B-29 wing, the jet fighter's power was reduced to idle or to engine cutoff.

As with the earlier Q-14 hookups, the F-84s ran out of aileron authority when trying to stay under control while attached to the B-29 wingtips. And, as had been learned with the Q-14, the application of elevator could provide the additional control input necessary for stabilizing the attached F-84. However, a problem arose because the locked-down F-84 was pitch-limited, and the application of elevator could induce twisting in the B-29 mothership's wing.

Measurement of the twist phenomenon satisfied the testers that it was not excessive. The left-hand F-84 began making hookups on the third flight of Tip Tow. As problems were rectified, dual towed flight took place with both F-84s locked down on the B-29's wingtips on flight 10 on 15 September 1950. The longest dual towing flight lasted more than 2.5 hours.[36]

The wingtip-mounted fighter-towing concept reached its ultimate expression in the Tom-Tom B-36 experiment. Although dry hookups were accomplished, the endeavor was far from developed. A serious rolling incident experienced in the connected F-84 on 26 September 1956 led to wingtip damage on both aircraft and an end to the program. Ever more effective aerial tanker aircraft enabled ever more capable jet fighters to cover long distances instead. (AFFTC/HO)

Anderson summed up the lessons of Tip Tow: "Towing appears feasible in smooth air. The concept was validated with quantitative data showing that a bomber could tow two fighters on its wingtips with a very small degradation in range. From the performance data gathered during the two long dual towed flights, a drag polar curve was generated . . . [and] plotted along with a similar curve for the B-29 flying alone. This shows that at the higher gross weight and lower airspeeds, the B-29/F-84 combination is more efficient than the B-29 alone. However, at low gross weight and high airspeed, the B-29 alone has the best performance characteristics. The floating panel concept is certainly validated. . . ."[37]

The Tip Tow aircraft subsequently underwent modification in an effort to achieve automatic flight control instead of manual operation by the fighter pilots once attached and locked to the B-29 mothership. On 24 April 1953, the left-hand F-84 was coupled by itself to the B-29. Momentary activation of the automatic flight control system from

Chapter 2: Parasites with a Purpose

the fighter quickly yielded a disastrous inward roll of the attached F-84. The B-29's outer wing panel began to crumple, and then the F-84, now inverted over the B-29 wing, struck the bomber's wing spar. The F-84 lost its nose forward of the cockpit; both test airplanes separated, coming down out of control. The spiraling B-29 mothership plunged into Peconic Bay, off eastern Long Island. There were no survivors in either aircraft.[38]

Tom-Tom Project

Tom-Tom was another effort to hook a pair of F-84s to the wingtips of a giant B-36. Envisioned were connectors that would allow the F-84s to replenish their fuel from the B-36 in flight while hitching a ride. Convair conducted the program in Fort Worth, Texas. Thieblot Aircraft Company of Bethesda, Maryland, as subcontractor to Convair-Fort Worth, was awarded a contract for the engineering study, design, and development of the wingtip coupling system. Tom-Tom was initiated to provide a weapon system for existing requirements of the Strategic Air Command. The need existed for a maneuverable fighter aircraft capable of long-range operations for either reconnaissance or bombing missions.

Tom-Tom, consisting of a B-36 carrier and two RF-84F parasite fighters, was one solution to this need. The RB-36F used for Tom-Tom was number 49-2707, the same bomber as the FICON test mothership. The two swept-wing RF-84Fs earmarked for the tests were 51-1848 and 51-1849. Although Tom-Tom sought the safe carriage of two RF-84Fs on the wingtips of a B-36 mothership, planners already looked ahead to the possibility of combining Tom-Tom with FICON, in which one RF-84 would be carried partially exposed in the bomb bay of a special B-36. This would create a super B-36 mothership capable of carrying three RF-84s simultaneously.[39]

Building on the work previously done by Tip Tow and the early C-47/Q-14 tests, the Tom-Tom team had its work cut out for them. Acknowledged hurdles to be overcome included automatic control of the towed aircraft, coupling to the mothership in turbulent air, and the employment of a quick, positive release of the towed fighter triggered by a predetermined angle of towed aircraft roll, or flapping. By November 1954, the Strategic Air Command no longer had a need for Tom-Tom; changes in tactics and newer aircraft designs rendered it obsolete. But research with the Tom-Tom test aircraft continued, to pursue the concept of a floating wingtip as a range extender for aircraft. The use of low aspect ratio wings on supersonic aircraft designs of the 1950s might achieve greater range and greater subsonic flight efficiencies with the use of jettisonable outer wing panels that also served as fuel stores, while increasing aspect ratio.[40]

The Tom-Tom team concluded that a type of wingtip hookup called a skewed axis coupling could render the F-84 stable in its connected position, removing the need for autopilot inventions, thereby simplifying the problem. A two-point locking latch held the RF-84 wingtip in such a way that the RF-84 could not move in yaw or traditional pitch. Its only freedom of movement was in a skewed line cause by the location of the mounts that allowed the RF-84 to roll, albeit not on its normal thrust axis.

Following wind tunnel test results, this setup appeared to work on a number of fully locked-on flights with a single RF-84F attached in flight to the RB-36F mothership. To achieve this locked-down attachment, first the RF-84 pilot had to engage the front latch by operating a pair of vertically opening jaws on the RF-84 wingtip that captured a coupling head on the B-36 wingtip. With only the front latch engaged, the RF-84 still had the ability to move in pitch, yaw, and roll about the connector. The engaged front latch retracted to a coupled flight position where a sequence of toggles and mechanical actions locked down the RF-84 in pitch and yaw at the rear attach point, leaving only the skewed axis roll function moveable.

The two-point latch mechanism featured an automatic release that was supposed to separate the aircraft any time the RF-84 moved in roll more than a predetermined number of degrees from normal flight attitude. The ultimate plans, which were never executed, called for utilities to be connected to the RF-84 once it was locked onto the RB-36 wingtip. The connections were to provide for fuel, air, electrical power, and interphone. This would enable the RF-84F to refuel while on the wingtip mount, and it would enable the RF-84 to shut down its engine, with ancillary power and pressurization capability available from the B-36. Had Tom-Tom been developed further, inlet-closing doors for the RF-84Fs were envisioned to stop engine-windmilling drag when the jet was riding without its engine lit.[41]

The basic RF-84Fs required wing modifications on one side to accommodate the coupling mechanism. The mothership RB-36F also required a new wingtip mount structure. A new spar was added between the aft attach fitting of the bomber's coupling device, running inboard through the trailing edge to wing bulkhead 38. This spar and its cover plate stringer provided additional strength. But mostly it increased torsional rigidity, a Tom-Tom summary report noted. Beefed-up and new bulkheads in the B-36 wingtip were based on anticipated loads imposed by mating an RF-84 with the wingtip. The whole wingtip mount on the B-36 changed the look of the previously rounded B-36 wingtip. Special lights were installed on both aircraft to facilitate night couplings.

With Strategic Air Command's withdrawal from Tom-Tom based on mission changes, the program became a pure research effort. Funding was tight, but the Tom-Tom team continued to work as much as it could to perfect wingtip carriage of aircraft.

Before actual inflight coupling was accomplished, one of the RF-84Fs had a mock-up wingtip modification made on its right wingtip; the intended B-36 coupling device was simulated on the test B-36's left wingtip. This allowed approaches to be flown within inches of coupling to test the ability of the RF-84 pilot to make a connection.

The test team was vitally interested to learn about turbulence and vortex action at the big B-36's wingtip. The RF-84 flew approaches to the modified B-36 wingtip as well as to a standard clean B-36 wingtip. Turbulence behind the modified wingtip with the coupling mechanism mock-up was more pronounced than was the turbulence behind a clean wing. Nevertheless, vortex effects behind the modified B-36 mothership wing were less than those measured behind a standard clean wingtip.

The complex interactions of airflow caused by the proximity of two aircraft in flight were examined for Tom-Tom. The B-36, even with the modified wingtip, produced a noticeable vortex as far as 30 feet behind the B-36's wingtip. Judicious use of throttle and considerable aileron inputs from the RF-84 pilot were required to penetrate the vortex area.

On the first flight where contact between the two aircraft was accomplished, the functional coupling hardware was installed on the right wingtip of the B-36 and the left wingtip of the RF-84F. Once the RF-84F penetrated the trailing vortex area, testers noted that the RB-36 tended to yaw left. This could be controlled, but it demonstrated the interactions between parasite and mothership.

In early June 1956, the first unofficial two-point hookup occurred; the yaw lock was not engaged, but the RF-84 pilot made control inputs that enabled this quasi-hookup to continue for about 23 minutes. The 12th flight of the Tom-Tom program on 8 June 1956 saw the first full hookup, generating a 45-minute connected flight. After this, the program stood down for lack of funds until 31 August, when more coupling flights resumed, including one with a duration of 1 hour and 20 minutes. The concept of skewed axis looked promising.

On 26 September 1956 Jordan and Israel clashed in the Middle East, Dwight Eisenhower was president of the United States, and the space race had not yet begun. Over north central Texas, a flight-test drama played out. On Tom-Tom's 18th flight, with one F-84, the combination aircraft showed the likelihood of reaching airspeeds where neutral stability or even instability could occur.

On a single-point contact, the RF-84F abruptly started rolling oscillations, increasing in amplitude. By the third cycle, the RF-84F was 20-degrees wing up. The pilot experienced severe buffeting that hindered control. The RF-84 rolled down, with the pilot starting the rapid-release process. The RF-84 was anywhere from 30- to 45-degrees wing down when the separation occurred, accompanied by shuddering in the B-36 airframe.

Things happened quickly, and testers never determined if the separation of the two aircraft was because the RF-84 pilot initiated it or because the rolling RF-84 prompted the automatic roll-release mechanism. Portions of the RB-36's coupling mechanism fell to earth. Both the RB-36 and RF-84 made safe landings, where inspection showed wingtip damage to both aircraft along with coupling equipment damage.

At the urging of Convair-Fort Worth, all future Tom-Tom test flights were canceled. Subsequent data analysis confirmed the wisdom of this choice. The official ending of Tom-Tom was received in a termination letter from the Air Force dated 28 March 1957. All special instrumentation was removed from the two Tom-Tom RF-84s and their mothership B-36. The B-36 made a one-way ferry flight to Davis-Monthan Air Force Base; the two RF-84s were shipped to Hill Air Force Base, home of the F-84 depot program.[42]

Chapter 3
BOMBER BONUS TESTBEDS

EB-45C 48-009 is shown banked to reveal the GE J79 jet engine mounted ventrally for flight testing. The J79 famously powered the F-104, B-58, F-4, and other aircraft. (General Electric)

Although the use of existing airframes as motherships and testbeds is nearly as old as flight itself, the concept flourished in the United States after World War II when the country was awash in retired, obsolete, and relatively low-time bombers from that war.

Beating Swords Into Plowshares

The Army Air Forces (AAF) leadership, including General Henry "Hap" Arnold, remembered what happened after World War I, when a surfeit of U.S.-built DH-4 and JN-4 biplanes led a frugal Congress to presume the service's needs could be met with these vintage airframes. The lingering use of DH-4s, carving into the precious funds for new research and development of state-of-the-art designs, stagnated military aviation. That there was any development and growth in the 1920s may be attributed to shrewd husbanding of resources and focused goals for aviation development in the face of fiscal austerity.

Correspondence to and from Hap Arnold during World War II showed that he was acutely aware of the risk of

Obsolete B-17G Flying Fortresses, some of them apparently never issued to a unit, were demolished with dynamite at Holzkirchen, Germany, when photographed on 22 May 1948. The rapid removal of so many bombers from the post-war Air Force may have facilitated the use of these types in the United States for test purposes, so the Air Force could concentrate on acquiring new bombers to meet new threats. (AFHRA)

In April 1947, the last seven B-24s to return from a wartime theater flew from Alaska to the lower 48 states. Five months later, in the face of newer, larger bombers being built and planned for the Air Force, the B-24 was unceremoniously reclassified as a Light Bomber instead of the Heavy Bomber it had been throughout the war.[43] Clearly, any available B-24 airframes could be expended as testbeds by that time.

General Electric Tested with Purpose

The inventive genius pool resident at General Electric in upstate New York embraced aviation as early as 1918, when a team led by Frank Moss demonstrated a viable turbosupercharger. But they had no testbed aircraft capable of taking the device safely aloft initially, so Moss and his team made a logical substitution and trucked a Liberty engine equipped with a turbosupercharger to the 14,000-foot level at Pikes Peak in Colorado.

The Liberty's output is listed between 400 and 425 hp at lower levels in denser air. In the atmosphere at 14,000 feet, a normally aspirated Liberty engine's horsepower dropped to only 230. Moss' terrestrial-bound Liberty with a turbosupercharger is said to have achieved 350 hp at altitude. That was enough to gain positive attention from the young Air Service. In a sense, perhaps the first testbed aircraft in America were a few LePere LUSAC-11 fighters fitted with the new turbosupercharged engine beginning in 1920.

During the 1920s General Electric (GE) hired an airplane and pilot at the Schenectady, New York, airport to flight test radio equipment, autopilots, and other systems under development by the company at the time. By 1931, General Electric entered the air age by purchasing a Stinson Model W cabin monoplane that alternated between test aircraft and executive transportation for the company.

As the 1930s progressed toward the wartime 1940s, GE's aviation systems developments became increasingly sophisticated, demanding ever more flight testing. Work

owning fleets of yesterday's airplanes once World War II was won. He was more than willing to load up war-weary heavy bombers and send them unpiloted over Germany in a pell-mell expenditure of old airframes that might strike targets of worth. Although such a wide-scale destruction of B-17s as guided bombs did not follow the more precise testing of Aphrodite remotely piloted B-17s against German targets, it nonetheless telegraphed Arnold's concerns that a post-war Congress would once again curtail new bomber programs in favor of re-using existing aircraft.

The architects of the post-war U.S. Air Force expeditiously and unsentimentally scrapped fleets of warplanes, both in the United States as well as overseas locations including Germany. Initially, B-29 Superfortresses were retained in greater numbers since they were viable stopgap heavy bombers pending the operational availability of B-50s, B-36s, and B-47s. B-17s, long the docile darling of the generals for transportation, remained in non-bomber service in limited numbers while B-24s faded from the Air Force en masse, with but a handful in service as bombers into 1947.

was initially contracted out. In 1941, General Electric was contracted to develop turbosuperchargers for the new B-29 Superfortress program. In addition, General Electric developed similar remote turret guns systems for the B-29 as well as the Northrop P-61 Black Widow and Douglas A-26 Invader. The need for an in-house GE aviation department was apparent.

General Electric received on bailment a pre-war twin-engine Douglas B-23 (39-050) bomber for the B-29 turbosupercharger testbed and a pair of B-24C Liberator bombers (40-2385 and 40-2386) for use as armament testbeds in November 1942 and May 1943.

A third B-24 (probably Ford B-24E 42-7057) joined the GE bailment air force in June 1943. Only nine B-24Cs were built before production shifted to the first Liberator model to see combat, the mass-produced B-24D. The C-models were not destined for combat, so their use as testbeds was a frugal decision. A Convair flight instruction manual covering early models including the handful of B-24Cs described the unique armaments carried by B-24C 40-2386: "The gun installation for this airplane consists of one .50 caliber machine gun in the nose, two .50 caliber machine guns in the upper electrically driven turret, two .50 caliber machine guns in the bottom non-retractable, electrically driven turret and two .50 caliber machine guns in the electrically driven tail turret. All turrets are General Electric.[44]

The early Ford B-24E likely also lacked combat suitability; many E-models were retained as stateside trainers, and again this airframe's bailment for testbed purposes is logical.

As General Electric was still growing and maturing as an aviation user, it was appropriate to contract with American Airlines to perform modification and maintenance services for the bailed bombers at American's La Guardia Airport base in New York City.

In anticipation of its upcoming duties on the B-29 turbosupercharger program, the olive-and-gray B-23 was heavily modified for its role as a testbed aircraft. Outsized four-blade propellers were fitted, and the engine nacelles were extensively redesigned and rebuilt to accommodate turbosupercharger installation. The cabin was modified to accommodate pressurization. This rework began in early 1942 and was not finished until that December.

Using one type of aircraft to host components being tested for another can lead to unintended consequences. The GE B-23 exhibited buffeting at high speeds caused by the altered wake after the nacelles were reshaped. Further changes to the nacelle shapes helped diminish the buffeting, but it never left entirely. Even though the XB-29 flew before the testbed B-23 was airborne, this turbosupercharger testbed nonetheless contributed to the maturation of the B-29 program. The strange B-23 with four-blade propellers experienced a hydraulics failure in flight that necessitated an emergency landing without benefit of brakes or flaps; some airframe damage resulted, but the airplane was repaired.

In late 1943, General Electric switched from American Airlines at La Guardia as the support contractor for aviation to Pan American World Airways, which had a base in warm Brownsville, Texas. The move to Texas was a mixed blessing. Brownsville could be a better locale for flight

This is likely a photo of the GE remote turret tested on testbed B-24C 40-2386. This turret does not appear to be retractable similar to the Bendix and Sperry lower turrets associated with some B-24D and later Liberators. (AAF)

34   Testbeds, Motherships & Parasites

testing, but it was so deep in Texas that its distance from the rest of the GE establishment in New York was vexing. The B-23 and a pair of B-24s migrated to Brownsville in the spring of 1944. Some work on the electric turret system intended for the B-29 was accomplished on the B-24 testbeds at Brownsville, but in August 1944, a GE/Pan American crew went to Wichita, Kansas, to pick up a brand-new B-29 (42-24716) to complete the testing.

Concurrent with GE's movement of aviation operations to Brownsville in late 1943 to early 1944, an additional Ford-built B-24H Liberator (42-95100; sometimes called an XB-24J) was delivered to Brownsville (see below). Pan American undertook modifications to make this Liberator into the testbed for the GE I-40 jet engine. It was not the last Liberator to mount a jet engine.

A second B-29 Superfortress (42-63571) delivered to Brownsville circa late 1944 was modified there to test the intended tail gun system for the unbuilt Convair XB-36. Radar of the type to be installed in the B-36 was grafted onto the aft fuselage in a pod offset to the pilot's side of the fuselage, beside the B-29's original gunner's compartment.

XB-29G GE Testbed

The designation XB-29G was applied to a single B-29 (44-84043) used by General Electric to place a jet engine in flight conditions. The jet rode in the Superfortress' bomb bay until test conditions were reached, at which time it was extended beneath the bomber into clear air at flight speeds that could provide useful test data not attainable in static ground engine test stands.

Aircraft number 44-84043 was GE's third B-29. It was part of the Bell production line at Marietta, Georgia. It arrived in Brownsville late in 1945 almost new, after receiving modifications in Georgia to enable it as a testbed aircraft for the TG-180 axial-flow jet engine. The TG-180 lent experience to General Electric in the development of the TG-190, designated J47, one of the most successful early jet powerplants.

During flight tests of GE's J73 engine, the testbed B-29 was able to shut down all four piston engines and maintain a speed of 180 mph, powered only by the one jet engine suspended from its bomb bay. The J73 was chosen to power the F-86H model of the Sabre. The J73 was a development of the earlier highly successful J47 jet engine.

The GE rationale for airborne jet engine testbeds was premised on three distinct phases of testing a jet engine. Early in the development of a jet engine, with unknowns to be expected, it was appropriate to host the unknown powerplant in a testbed aircraft that could fly on its own engines, not relying on the experimental jet under test. Then, once the basic engine performance was validated and any problems overcome, it was useful to place the engine in a special wind tunnel that could simulate altitude, speed, and temperature conditions beyond the realm of the actual testbed aircraft. Complete performance evaluations could be made on the engine at a time when it might be ready for production, or at least be beyond the early teething problem stage of development.

The NACA's altitude wind tunnel in Cleveland was the premier facility for this work. Once this data was combined with the earlier experiences with the engine, it was finally appropriate to fly the jet engine in the aircraft for which it had been designed.

Liberators Light the Fire

Although the Air Force battled mightily for the B-36 in the late 1940s, the successful dismantling and discrediting of B-17 Flying Fortresses and B-24 Liberators as bombers made these airframes available for odd purposes. As far back as the wartime era, General Electric saw the promise of B-24 Liberators as jet engine testbed aircraft. The general layout of the boxy, high-wing B-24 permitted a turbojet powerplant to be installed in the waist section, with its inlet ducted from the top of the fuselage. The hot exhaust exited through the aft fuselage where the tail turret had been.

Anecdotally, the March 1942 flight testing of a German Jumo 004 axial-flow jet engine mounted on a Messerschmitt Me 110 twin-engine fighter may be the first use of a multi-engine airplane as a host testbed to flight test a jet engine. At about the same time, a British Wellington bomber carried a Whittle jet for testing, its exhaust emanating from the rear of the Wellington's fuselage.

Documentation indicates at least four B-24 Liberators were modified in this general format to test early turbojet engines. One of these is a double curiosity. It shows up on the Convair San Diego ramp in September 1946 with a dorsal inlet, and it carries the serial number of a twin-tail

This converted B-24J (42-73215) was photographed in 1946 with a dorsal jet inlet as well as a nonstandard single tail configuration. (Convair)

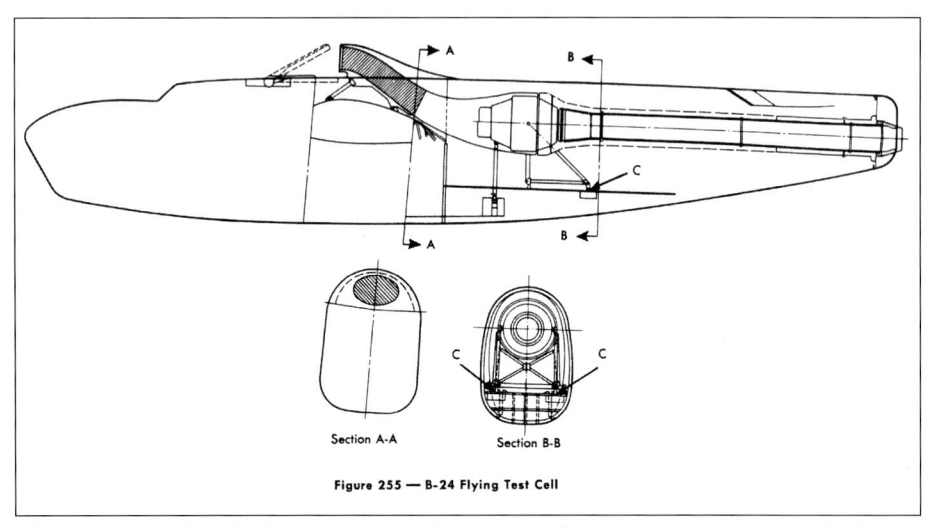

This 1945 GE sketch depicts a simplified view of a turbojet testbed made from a B-24 Liberator bomber. A new dorsal inlet for the jet engine was ducted over the wing carry-through area and back to the jet, mounted in the waist section of the B-24 in a way that allowed thrust measurements to be taken. Long exhaust vents were placed through the bomber's former tail gun turret location.

Liberator on a newly modified single tail. This testbed was B-24J 42-73215, modified to carry a GE J35 jet engine for test work supporting Convair's post-war B-46 jet bomber.

Line art from a 1945 GE pamphlet depicts a greenhouse-nose early model B-24 fitted with a jet engine in its aft fuselage. It is possible that a noncombatant B-24C or E-model may have been an early jet testbed, since General Electric had two C-model Liberators available at Brownsville, or perhaps the drawing was simply an available expedient.

The Liberator 42-95100 started life as a Ford-built B-24H. It was fitted to be that company's first B-24J and is sometimes known as XB-24J. It soon came to General Electric where it hosted the I-40 jet engine for tests before war's end, possibly as early as late 1944. By November 1945, General Electric vacated its Texas flight facility in favor of the company's Schenectady, New York, location.

An important part of the early jet testbed testing was the ability to place the jet engine at high altitude where restarts were more difficult. Testers could quantify restart performance and make changes to the design if necessary, while the jet engine was hitching a ride inside a B-24, before the engine became the primary powerplant for an aircraft. Early turbojets exhibited a tendency to "blow out" or have combustion burners cease functioning in idle throttle settings that occurred at higher airspeeds when high altitude was introduced. It was vital to establish these performance parameters and to comprehend possible improvements that could be learned from flying testbeds. A typical

The NACA used this San Diego–built B-24M (44-41986) as an engine testbed with a turbojet mounted in the waist section fed by a special dorsal inlet when this photo was taken at 7,000 feet over Lake Erie in 1946. A waterspray rig with nozzles is visible just ahead of the inlet. The exhaust section extended from the lower portion of the tail turret. This B-24M later mounted a Westinghouse jet on a wing pylon next to the right fuselage and received a radically modified nose for ice accretion testing. (NACA)

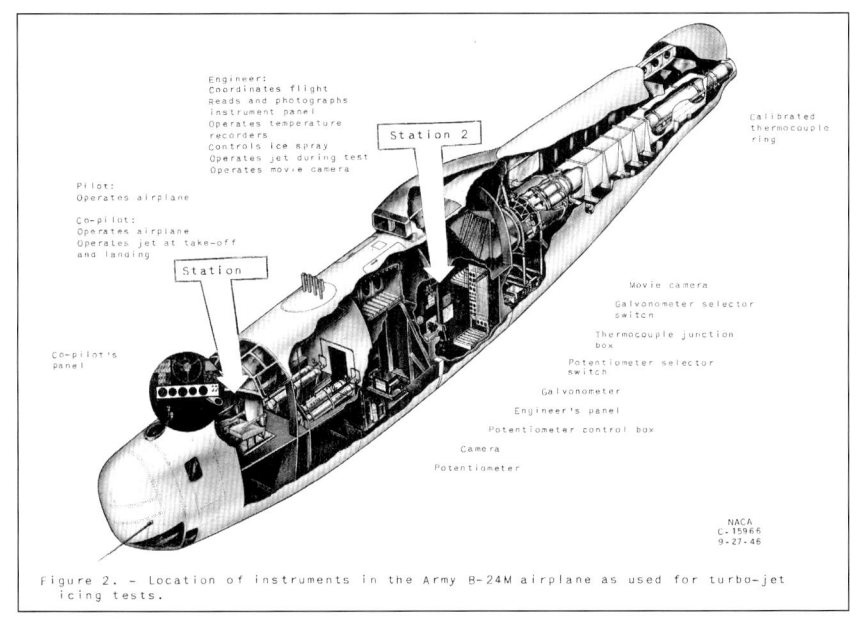

This NACA cutaway artwork shows the interior of the modified B-24M jet engine testbed as it looked in 1946. (NACA/NASA Glenn Research Center)

This view of the test engineer's control panel in the NACA B-24M looks aft along the left side of the fuselage in the 1946 configuration, which featured a turbojet mounted in the waist section. (NACA)

B-24 waist compartment turbojet test mount might include pneumatic pistons instrumented to allow engine thrust measurements to be taken.

B-24 Liberator testbeds also appealed to the NACA. The B-24 allowed the jet engine to operate in flight conditions at various altitudes; the four Pratt & Whitney R-1830 piston engines provided actual power (and redundancy) for the B-24. The jet engine could be flown and not imperil itself or the test crew because the former bomber's original engines kept it aloft. B-24M (44-41986) hosted a jet engine in its waist section shortly after the end of the war. The NACA conducted early jet engine icing and water ingestion tests by placing nozzles just ahead of the dorsal inlet and

Chapter 3: Bomber Bonus Testbeds   37

These rulers measure the ice buildup remaining on the test turbojet after the NACA B-24M landed. (NACA)

The NACA sought out punishing inflight icing conditions for the Westinghouse 24C turbojet engine mounted beneath the wing of its B-24M in Ohio on 15 March 1948. (NARA/NACA)

The NACA B-24M with a Westinghouse jet engine is bedded down on a snowy ramp at the Lewis Flight Propulsion Laboratory in Ohio in 1948. (NARA)

38   Testbeds, Motherships & Parasites

controlling the spray flow intended to see what jets would do in such conditions.

Later, this same NACA B-24 engine testbed used a simpler approach, mounting a Westinghouse 24C-2 jet engine on a pylon beneath the right wing for free-air icing testing. The pylon was far enough inboard and rode low, which kept the jet engine's inlet out of direct line to the B-24's number-3 engine prop wash. Jet inlet icing was considered a serious hazard to be studied, understood, and counteracted in the immediate post-war years.

The NACA B-24 testbed's first hour in real-world icing conditions during a March 1948 test saw marked changes

The camera and lighting pod sat ahead of the 24C jet engine on NACA B-24M, shown here at the NACA Lewis Flight Propulsion Laboratory near Cleveland, Ohio. (NARA/NACA)

The NACA mounted a Westinghouse 24C jet engine on a vertical pylon from the right wing inboard of the engines. A braced pod ahead of the engine mounted a camera and lighting. This engine was developed as the J-34. (NARA/NACA)

B-24M 44-41986 was a NACA testbed after the war, carrying a Westinghouse 24C turbojet on a pylon beneath the right wing when photographed. This aircraft also performed windscreen icing tests for the NACA. (NARA/NACA)

Chapter 3: Bomber Bonus Testbeds 39

Close-up photo shows the mounting of a Westinghouse jet engine on NACA B-24M in Ohio in 1948. The unusual nose configuration is an artifact from other icing tests performed by the NACA. (NARA/NACA)

The camera and light pod in front of the turbojet on NACA B-24M 44-41986 could be raised or lowered. (NARA/NACA)

to the tested jet engine, turning at a speed of 9,000 rpm. After 45 minutes, the jet engine's exhaust temperature had increased at the tailpipe from 761 to 1,065 degrees F while thrust had diminished from 1,234 to only 910 pounds. With these performance statistics, the test engine did not have to be shut down, but a reduction in jet engine speed would have been necessary had the jet been running at the takeoff power rating of 12,000 rpm.

The following month, the B-24 with the jet engine riding beneath its wing found icing conditions again, this time while the turbojet was spinning at a normal cruising speed rotation of 11,000 rpm. Engine thrust dropped noticeably as the Liberator flew into the ice-producing cloud. Ice quickly added layers to the inlet lip, disturbing the airflow and degrading pressure recovery for the jet engine. Subsequent changes in atmospheric conditions changed the jet's performance. Although this test could not yield general conclusions about jet engines and icing, it did confirm the potential seriousness of the problem, while using a safe four-engine testbed bomber as host for the lone turbojet. This same NACA B-24M also conducted windscreen de-icing testing.

General Electric sent another B-24 (44-49916), a Ford-built L-model, to American Airlines at LaGuardia for modification to carry the proposed B-47 Stratojet tail gun system and radar. As such, it was known as the XB-24Q.

These late-war projects by General Electric occurred over a backdrop of activity as the company made plans by May of 1945 to build its own flight test facility in Schenectady. In keeping with its role as a leader of industry, General Electric made the new Schenectady hangar a striking concrete-arched structure, with hot water piped under the floor heating that surface against New York winters. C. G. Talbot managed the Flight Test Division for General Electric at Schenectady, staffing the operation with GE engineers and technicians while initially contracting with American Airlines for flight crews. Talbot also kept GE's corporate executive flight operations close at hand at Schenectady.

Post-war Liberators were among the first aircraft to populate the new Schenectady site, with the XB-47 tail gun testbed (XB-24Q 44-49916) taking up residence outside the unfinished hangar, followed in the spring of 1946 by the jet testbed B-24H 42-95100. That April, B-29 44-84043 cruised to Schenectady.

B-24 42-95100 never flew again. GE testing of the I-40 or J33 jet engine was finished, and the company based the next generations of jet engine tests on larger B-29 Superfortress bomber testbeds. Accordingly, the Liberator that had once been called Ford's XB-24J was returned to the Air Force. With little use for this now-obsolete bomber that had been extensively modified for its jet engine testbed role, the Liberator

A Ford-built B-24L Liberator (44-49916) was modified with a test radar and 20mm gun emplacement in the tail for the B-47 program. Designated the XB-24Q it was operated by General Electric. (Peter M. Bowers Collection)

was scrapped at Schenectady. Its Hamilton Standard propellers and Pratt & Whitney R-1830 engines were removed for possible future use. The airframe was broken up, literally, with demolition charges.

The early post-war era was marked by slow periods for some contractors. With victory, vast wartime efforts were abruptly canceled, and the future cadence of the Cold War was not yet driving research and development. GE's commitment to a new flight test facility in this economic climate was a testament to the company's faith in the future of such work. In 1946 and well into 1947, the two GE flight test projects of greatest significance using testbed aircraft were the XB-24Q (B-24L) with B-47 tail armament and the B-29 that tested the J47 engine suspended beneath its bomb bay.

The B-29 44-84043 was a special airframe. When new at the wartime Bell plant in Marietta, Georgia, it was earmarked for use by General Electric. From June to November 1945, this B-29 remained at a modification center in Georgia where it received bomb bay structure for the express purpose of carrying and testing the GE-designed TG-180 axial-flow jet engine. Allison later put this powerplant into production as the J35. It powered the F-84 Thunderjet, F-89 Scorpion, as well as a variety of experimental and low-production aircraft of the immediate post-war era.

The Boeing XB-47 Stratojet first flew with J35 engines, quickly switching to more powerful J47s also designed by General Electric. This Bell-built B-29, always to be a testbed, acquired the nomenclature of XB-29G. It was the sole G-model created. The XB-29G began its working testbed career at Brownsville in December 1945 with a Pan American World Airways flight crew and GE technical crew on board.

Initial tests of the TG-180 jet engine were made with the powerplant neatly cowled in a nacelle that lowered on a trapeze from the bomb bay in flight. To accommodate this, the XB-29G's aft bomb bay doors were removed. But exigencies of flight testing, seemingly always on a frantic time schedule, made the nacelle more baffle than beauty, and some later photos of the XB-29 in flight show an uncowled jet engine to provide technicians quicker access on the ground.

The center main fuel tank of the modified XB-29G carried jet fuel for the test engine. In a typical B-29, the forward fuselage cockpit area is pressurized. A small circular tube provides a pressurized crawlway over the unpressurized bomb bays, leading to another pressurized crew compartment in the aft fuselage traditionally occupied by gunners. For GE's XB-29G testbed, controls for the jet engine and instruments for monitoring its performance were placed in the aft pressurized compartment. The typical GE XB-29G test crew, not including flight crew, consisted of a test conductor who operated the jet engine, a test engine crew chief, and two engineers who monitored instrumentation and served as scanners on either side of the Superfortress.

General Electric designed the TG-190, later to become the famous J47, at company expense. It was within the size parameters of the TG-180 but promised better performance. For testing in 1948, the XB-29G was easily adapted to the new jet. Test protocol included a ground run of the mounted jet engine before the XB-29G took off for altitude. To enable this, the B-29 straddled a special open-ended concrete trough, which gave clearance for the jet to extend as well as unimpeded exit flow for the jet's exhaust gases. However, the operation of the jet in such close proximity to the ground caused excruciatingly loud jet noise in the aft fuselage of the B-29 and cabin temperatures that could exceed 100 degrees F.

The J47 jet engine test program was GE's most ambitious effort in that arena at the time. In a program that spanned more than four years, two B-45 jet bombers could augment the B-29, if needed. Even as the J47 found favor in the sleek Boeing B-47 Stratojet and the aggressive F-86 Sabre jet fighter, this jet engine added unique benefits to the performance of the B-36 strategic bomber. The B-36 originally performed its duties powered by six Pratt & Whitney R-4360 piston engines driving pusher propellers. For extra power as needed, the B-36D and subsequent models added four J47 jet engines in paired pods under each wing. Unlike most J47s, the B-36 version ran on avgas instead of jet fuel.

The B-36 J47s needed to be able to start and shut down at high altitude for extra power as needed, something not normal to the flight regimen of most J47s. Such high-altitude activity could cause the jets to exhibit adverse startup traits above 30,000 feet. These problems were fixed but could not have been divined without the ability to fly a J47 on a testbed up to that altitude.

On occasion, with the J47 engine and later the more powerful J73, the test crew made flybys in the XB-29G with all four R-3350 piston engines shut down, to the astonishment of onlookers.

The career of XB-29G 44-84043 came to an end on 27 July 1953 while undergoing maintenance in the Schenectady hangar. An electrical short led to a hydraulic fire as workers scrambled from the B-29. The fire crew tried foam to smother the blaze, but the bomber continued to burn on the hangar floor. A tug tried to pull the XB-29G out of the hangar only to be thwarted by the slippery foam covering the concrete floor. A second tug was quickly attached to the first, and the hopelessly burning bomber was pulled onto the ramp, sparing other assets in the hangar.

B-50 Replaces XB-29G

The loss of testbed XB-29G 44-84043 was mitigated by the bailment of a B-50A (47-113) by the Air Force to General Electric. Structural modifications were made by the Air Force at Kelly Air Force Base in San Antonio, Texas. The bomb bay engine mount and retraction mechanism for the B-50 was designed to accommodate an extended tailpipe for J73 engine afterburner tests (known as TPA, tailpipe augmentation). Some test equipment salvaged from the burned XB-29G was installed in the B-50, saving time and money.

The B-50 arrived at Schenectady in the summer of 1954, almost a year after the loss of the XB-29G.

GE's B-45 Testbeds

General Electric took advantage of the all-jet B-45's long 22-foot bomb bay (sufficiently large for nuclear weapons

At Schenectady, General Electric devised an open-ended trench to allow jet engines suspended from bomb bays to be lowered for work and ground testing, their exhaust flowing out the back of the trench. Poised over the trench in this photo is a North American B-45 with a lowered J79, probably sometime in 1955. (General Electric)

42   Testbeds, Motherships & Parasites

delivery) to make another mount for airborne jet engine testing. The B-45 was an interesting stopgap jet bomber of the immediate post-war era, supplanted by more famous, and more numerous, nuclear-capable bombers, including the B-36, B-47, and B-52. The availability of two B-45s to General Electric in the 1950s increased the company's available airframes for in-flight testing of new jet engines.

In September 1949, General Electric provided specifications to the U.S. Air Force for a jet engine testbed based on the largest engine that the company contemplated at the time. The Air Force made B-45C 48-009 available, and North American Aviation, builder of the B-45, modified it to suit GE's requirements. By the spring of 1951, a North American pilot evaluated the testbed B-45 before turning it over to the Air Force, for delivery to General Electric.

Whereas some testbeds used trapeze mounts to swing the test engine into free air, this resulted in fore-and-aft motion during extension and retraction. The engines envisioned by General Electric would fill the B-45's bomb bay to an extent that vertical extension and retraction were necessary in the B-45 testbed; there was no room for swinging forward or rearward. After modification by North American, but before the B-45 arrived at Schenectady in 1951, it stopped at Edwards Air Force Base where the old loading pit for the original X-1 came in handy as a place to lower the B-45 bomb bay engine mount for inspection. General Electric also received the use of B-45C 48-010.

The former navigator's station in the nose of the B-45C was modified with a suite of jet engine test instruments for the GE test engineer. A periscope mounted in the nose gave

Boeing XB-29 (41-002) stayed with the company as The Flying Guinea Pig testbed. Here it mounts an Andy Gump nacelle on the number-2 engine. The photo is dated 26 April 1947 and was taken at Boeing Field in Seattle. (Boeing Photo Courtesy Tom Cole)

Chapter 3: Bomber Bonus Testbeds   43

North American EB-45A (47-096) carried Curtiss-Wright's J67 turbojet aloft circa 1954–1955. The B-45 bomber's contributions went beyond its intended operational service, providing a number of useful testbeds. (AFTC/HO)

the test engineer a view at the engine extended from the bomb bay.

A quirk of the jet engines lowered from the bomb bays of testbed aircraft such as the B-45 was their offset thrust. Operating so far below the normal thrust line of the host B-45, rapid application of power to the test engine could result in an unintended pitch-up of the aircraft unless the test engine operator and the pilot coordinated their actions.

GE B-45s test flew the company's J73 engine, mounted in a semi-retractable bomb bay sling starting in August 1954 and phasing out later that year. The J73 garnered contracts for the F-86H only.

Legendary among jet powerplants is the GE J79 engine. This axial-flow jet-propelled aircraft is as diverse and famous as the B-58 Hustler bomber, F-4 Phantom II fighter, F-104 Starfighter, and A-5 Vigilante. GE's B-45 48-009 was the sole engine testbed aircraft available at the end of 1954, and its modifications to receive the J79 carried into the spring of 1955. As flight testing continued, the GE B-45 J79 testbed had telemetering capability enabling it to transmit 12 channels of data to a ground station at the Schenectady complex.

B-45 48-009 took the J79 aloft for the first time on 20 May 1955 for its first inflight operation. The expanded performance envelope for the J79 meant that test flights in the B-45 had to breach 50,000 feet on occasion. Flight test crews wore pressure suits for these sorties. By the spring of 1957, the tests were finished and B-45 48-009 reverted to the Air Force.

B-26 Autopilot

General Electric employed a Douglas B-26 Invader with a long boom mounted over the nose to help devise a radar-autopilot-armament system. The boom extended into free air to record pitch and yaw to evaluate the system, circa 1953.

Around this time, General Electric was promoting the use of its test fleet of about eight aircraft for integrated testing. The idea was to use one airframe on one flight to record test data simultaneously on multiple projects that could include jet engine testing, armament resting, and avionics testing. The economies of time and cost were promising.

The 1953 GE test fleet available at Schenectady included a B-17, B-25, Douglas B-26, B-29, two B-45s, F-86, F-94, and F3D.

Westinghouse B-45

The capacious B-45 also became an engine testbed for Westinghouse. A JB-45A (47-049) served Westinghouse's jet engine developmental efforts.

NACA B-29 Engine Testbed

The effort to perfect the Boeing B-29 Superfortress and get it into combat in 1944, sometimes called the Battle of Kansas due to the site of Boeing's Wichita plant, involved the NACA as the country's agent of aeronautic developments. At the NACA's research operation near Cleveland, Ohio, the Aircraft Engine Research Laboratory took delivery of an early B-29 (42-6357) on 22 June 1944. The first B-29 combat mission had taken place a little more than two weeks earlier, but there were still performance items to improve to make the Superfortress the ultimate bomber of the war.

At Cleveland, the B-29 was evaluated for ways to improve the performance of its four hard-working R3350 radial engines. Overheating was a problem, and the Army Air Forces sought help from the NACA to rectify this. Reshaping of the B-29 engine cowling inlet and outlet (a specialty of the NACA for many years) aided Superfortress cooling while reducing drag. Improved fuel injection and cylinder head changes were devised. The NACA-derived improvements, in classic testbed style, were applied to only two of the B-29's four engines on the left wing, leaving standard powerplant packages on the right wing. Eleven test flights with military pilots confirmed the efficacy of the NACA improvements. The NACA results were available to the industry. A NACA wartime report on the topic does not clarify if these improvements were introduced in production, but their successful development on the B-29 testbed at Cleveland is known.

Mixed-Engine Mitchell

In the fast-paced whir of aeronautical developments during World War II, Hamilton Standard needed to test a propeller on different types of engines. The safety and redundancy to accomplish this test was accomplished by Pratt & Whitney specialists who took a B-25 Mitchell twin-engine bomber and replaced one of its Wright R2600 engines with a Pratt & Whitney R2800.[45] Hamilton Standard and Pratt & Whitney were often closely associated in propeller-engine combinations. Both companies were divisions of United Aircraft.

In the wartime year of 1944, the Pratt & Whitney enterprise at East Hartford used 18 types of aircraft for its many test requirements, logging an aggregate of well over 2,000 hours of flight time. Not all Pratt & Whitney test aircraft were multi-engine; a snapshot of the 1947 stable includes single-engine machines such as an F4U-4 Corsair, SB2U-1 Vindicator, Vultee AB-2, and Convair XA-41.[46]

B-17 Five-Engine Testbeds

If the Liberator lent itself to internally mounted turbojets in its waist, the circular cross-section of the B-17 Flying Fortress was employed in a strikingly different manner. On two occasions in 1946, Boeing modified surplus B-17Gs to house turboprop engines under development by Wright Aeronautical (Curtiss-Wright) and Pratt & Whitney. One was an Air Force aircraft initially bailed to Curtiss-Wright and modified at Boeing's Wichita facilities; the second was purchased on the market and modified in Seattle for Pratt & Whitney.[47]

A writer for Pratt & Whitney said that its civil-registered Fortress "was plucked off a war surplus scrap heap in Oklahoma."[48] The location was Altus, Oklahoma, where Pratt & Whitney bought two B-17Gs from the parked aircraft there in November 1947. One was ferried to the company's location at East Hartford, Connecticut, where it served as a hangar queen to keep the testbed B-17 in service. The other Fortress (44-85734) was given the civil registration N5111N (at one point NL5111N) and flown to Boeing's Seattle plant. Finished later than the Curtiss-Wright testbed, the Pratt Fortress was ferried from Seattle to Hartford with a rudimentary nose shroud in February 1949.[49]

Describing the Wright B-17 testbed in 1948, Boeing referred to the three-year-old B-17G that provided the basis of the testbed as an "elderly aircraft," albeit with respect for the aging warrior. Additional fuselage stringers were added for greater strength, as were four longerons. Heavier skin was applied to the beefed-up fuselage structure.[50]

Chapter 3: Bomber Bonus Testbeds   45

Pratt & Whitney's B-17G was briefly registered NL5111N before dropping the "L." It was taken into Boeing's Seattle plant complex for major rework into Model 299Z configuration, as seen here in October 1948. (The Boeing Company)

The B-17G initially bailed to Curtiss-Wright underwent Model 299Z rework at Boeing Wichita. Here, waist windows have been skinned over, extra stringers have been added, and a thicker skin will be applied to some of the fuselage to carry loads expected from the turboprop engine installation planned for this B-17 testbed. The cockpit has been removed in preparation for its relocation about 4 feet aft of its original location. (The Boeing Company)

Pratt & Whitney's five-engine B-17 was this modified B-17G-115-VE, which Boeing converted into an engine testbed at its Seattle plant in 1948. The pilot's side window behind the sliding portion is shorter as the fuselage was adjusted to take the new cockpit location. The nose shows new metal where engine cowling later resided, plus a minimally streamlined, pointed metal cap aerodynamically suitable for ferrying the special Fortress to Pratt & Whitney. (Gordon S. Williams Collection)

This left-side view shows the Pratt & Whitney 299Z at Boeing Field in Seattle in 1948, reinforced and reskinned for its new duties as a testbed for turboprops. (The Boeing Company)

In both cases, the turboprop engine rode in the nose inside a new cowling faired into the circular fuselage cross-section. Extreme measures were taken to provide enough space for the turboprop powerplants. The traditional B-17 cockpit was moved aft more than 48 inches (one account says 47 inches) to make room. As with the B-24 engine testbeds, the B-17s provided a margin of safety with their four traditional reciprocating engines. Nonetheless, testers seemed fond of showing the great power of the large turboprops, and images exist showing the B-17 testbeds flying with all four Wrights shut down, their propellers feathered.

It seems that Curtiss-Wright had its own notions about propellers for those Wright 1820s, installing rounded paddle propellers with Curtiss Electric hubs in place of the normal Hamilton Standards of the B-17 fleet. The Curtiss-Wright B-17 engine testbed housed the XT35 turbine powerplant capable of developing more than 5,000 hp. Pratt & Whitney's similar Fortress testbed mounted the XT34 propjet engine in the nose. After a period as an Air Force aircraft bailed to Curtiss-Wright, the five-engine B-17 built for the Wright XT35 was sold to Curtiss-Wright and given the civil registration number N6694C. Early test use of the Curtiss-Wright Fortress took place in New Jersey.

Removable ductwork carried the turboprop exhaust rearward and off to the left of the Pratt & Whitney 299Z five-engine testbed for ground runs. (The Boeing Company)

Chapter 3: Bomber Bonus Testbeds   47

The Wright Typhoon in the nose of the company's 299Z testbed B-17 propelled the bomber while its four original Wright Cyclone radial engines were shut down. Peculiar to the Curtiss-Wright test B-17 is the substitution of typical Hamilton Standard propellers with Curtiss-Wright props, rounder in shape and fitted with typical Curtiss hubs. (The Boeing Company)

The Pratt & Whitney T34 turboprop is seen uncovered in the nose of the company's 299Z testbed Fortress. Visible is the welded tube engine mount structure, and the exhaust ducting from the back of the engine out the bottom of the fuselage. The T34 made its first flight, in this B-17, in September 1950. (The Boeing Company)

The classic geometry of the B-17 Flying Fortress is unmistakable, even with a relocated cockpit and one powerful Pratt & Whitney T34 turboprop in the modified nose providing the sole means of propulsion. The T34 ultimately powered the four-engine Douglas C-133 transport. To accommodate the T34's ventral exhaust location on the testbed, the teardrop loop antenna had to be moved to a dorsal location. (The Boeing Company)

The Curtiss-Wright T49 propjet derivative of the J65 turbojet was tested on the nose of the five-engine B-17G used by Curtiss-Wright and later purchased by the company. The T49's propeller is feathered here on 13 August 1956. The T49 partially powered the experimental XB-47D. (AFFTC/HO)

The massive Wright T35 turboprop dwarfs the rest of the Wright 299Z five-engine testbed. Bifurcated ventral exhaust ducting is visible. The aircraft behind the B-17 is another Wright Aeronautical machine, a Lockheed Ventura NX33626. Earlier, this had been a Lockheed test aircraft. Its nose was shortened for propeller clearance and two engine, nacelle, and propeller assemblies for the C-69 Constellation were installed. (The Boeing Company)

The Curtiss-Wright B-17G (Boeing Model 299Z) briefly underwent further modification with a new underslung centerline mount for the J65 turbojet. Later, this testbed returned to its nose-mounted propeller engine configuration. (AFTC/HO)

Chapter 3: Bomber Bonus Testbeds   49

Wright R-3350 radial turbo-compound engine received testing on Curtiss-Wright's five-engine B-17 following its service on Air Force projects. (AFFTC/HO)

With a version of the R3350 reciprocating engine in the nose, the Curtiss-Wright 299Z variant of the B-17 was a popular curiosity at an Edwards Air Force Base open house in the mid-1950s. Following many years of testbed service, this B-17 was ultimately rebuilt with a stock Fortress nose and flew as a firefighting air tanker. (AFTC/HO)

The T35 did not lead to production. After T35 engine tests, this Fortress hosted a reciprocating turbo-compound version of the R-3350 engine. Turbo compounding used power recovery turbines to capture as much as 550 additional horsepower for takeoff and 240 in cruise from exhaust gases, as compared to the output of an earlier R-3350 engine. A chin modification also allowed this B-17 to test a J65 turbojet in a neatly faired nacelle beneath the forward fuselage. Curtiss-Wright bought rights to the British Sapphire

Fred S. Chamberlin, test director for the Wright Aeronautical Division of Curtiss-Wright at Edwards Air Force Base, was a former Marine pilot who was at the controls as the airplane made its last flight from Edwards on 27 December 1960. He said that the various engine modifications did not alter this airplane's original B-17 flying characteristics. (AFTC/HO)

jet engine, developed it, and sold it in the United States as the J65.

By 1956, the Curtiss-Wright B-17 engine testbed was at Edwards Air Force Base where it conducted test flights into 1960. The 10,000-hp T49 was one such powerplant. After the T49 tests, the Air Force sold the B-17 testbed to Curtiss-Wright. More turbo-compound R-3350 tests ensued.

The Curtiss-Wright B-17 was known for its reliability over many years of test work, although a pilot recalled one incident. During testing in New Jersey, the fifth engine's propeller went into flat pitch, creating so much drag that the four original Wright Cyclones could not keep the modified bomber in the air. Fortunately, and pragmatically, the testing was done close enough to home base that the B-17 could land safely. The Curtiss-Wright B-17 carried Air Force serial number 44-85813. After its service to Curtiss-Wright in the 1960s, this aircraft's unique hemispherical nose cap was re-attached and the aircraft was sold, ultimately becoming part of Arnold Kolb's Black Hills Aviation in Spearfish, South Dakota, where Kolb operated firefighting air tankers.

Kolb purchased a studio-prop B-17G fuselage that had been part of the television series *12 O'Clock High* and carefully spliced the stock B-17 forward fuselage onto the modified Curtiss-Wright airframe to create a more standard B-17 for firefighting. The nose job was performed on the open-air ramp at Spearfish. Kolb later recalled how the warming rays of the sun complicated alignment as the heated side of the fuselage expanded more than the shaded side, moving plumb bobs off-center until the cause of the irregularity was determined. N6694C flew as an air tanker until a mishap in North Carolina in 1980 burned the nose. The remnants were saved for possible inclusion in other B-17 restorations.

The Pratt & Whitney B-17 testbed primarily tested the T34 turboprop during its career as a five-engine aircraft. The T34, originally a Navy project, found use powering the Air Force's Douglas C-133 transport instead.

Pratt & Whitney's B-17 N5111N appeared to be out of a job by the early 1960s. It was parked and subsequently preserved in long-term storage. Late in 1964, Pratt & Whitney (United Aircraft) received a contract from the Bureau of Naval Weapons to conduct the preparatory stage of flight test, followed in the summer or 1965 by another contract to conduct 50-hour flight test programs on two propellers, 73EGB1and VC86260, accumulating a total of at least 100 flight hours on the B-17. For these tests, a GE T64 turbine, an engine that powered several helicopters and aircraft such as the Canadian de Havilland Buffalo twin-engine military transport, was mounted in the fifth engine position in the nose of the B-17.[51]

"When reactivated for this test, the aircraft had been in storage for two years and had not flown for five years," testers reported. It was necessary to take the B-17 out of preservation and apply new fabric to its control surfaces. New radio equipment was installed to meet Federal Aviation Administration (FAA) requirements. The FAA approved the refurbished B-17 testbed for daytime flight under VFR (visual flight rules). Flight test airspeeds ranged from 165 mph indicated airspeed (IAS) at 30,000 feet to 240 mph IAS at 5,000 feet.

"Maximum airspeed was established as the speed that could be maintained with the aircraft engines at maximum cruise power and the test engine at idle power," the testers explained. During the flight tests, the five-engine B-17 could achieve zero g-loading if needed, but the testers noted, "Attempts to run a negative g condition were discontinued when oil pressure was lost on all engines during this maneuver."

For these tests, the B-17 was flown briefly with its four primary Wright R1820 engines shut down, deriving power solely from the nose-mounted turbine and each of the two propeller installations. Tests included rolling the 299Z to 45 degrees in either direction of zero. The testbed B-17 was pitched as much as 25 degrees above and below its pitch axis while inducing 10-degree roll inputs right or left.[52]

The Pratt & Whitney 299Z testbed B-17 flew from 17 November 1965 to 6 May 1966, accumulating 120 hours and 40 minutes of flying time to gain the required 50 hours of test time on each propeller. The propellers passed their test and the revived N5111N was the testbed that made it feasible.[53]

In the following year, the Pratt & Whitney B-17 testbed passed to the Connecticut Aeronautical Historical Association (CAHA) in June 1967. Minus a fifth engine, the testbed Fortress nonetheless conveyed a sense of its Pratt & Whitney service while on outdoor display at CAHA's site near Hartford.[54]

What happened next was a natural disaster unusual for Connecticut and startling for its severity. A tornado touched

Pratt & Whitney's B-17 testbed was displayed with its fifth engine nose open and visible at the Connecticut Aeronautical Historical Association (CAHA) site near Hartford in April 1973. (Author)

down among the airplanes of CAHA's airpark on 3 October 1979, tossing aircraft and hurling a Grumman Albatross amphibian onto the B-17's fuselage. The remains of N5111N came to life again in the restoration of two B-17s combined to make one aircraft, flown on the warbird circuit as *Liberty Belle* until a forced landing and fire brought that effort to an end in 2011. Portions of this famous testbed may rise yet once again as the clamor for flying B-17s remains high.

A third B-17 turboprop testbed, serial 44-85747, was fitted with a nose-mounted Allison T-56 test engine while leaving the cockpit in its normal location. This B-17 came to Allison as a Navy PB-1 (BuNo 83999) for use in testing the XT-40 turboprop. The Air Force took charge of the airframe subsequently, as Allison used it for the T-56 program.

Still another B-17G, actually a Navy PB-1W (probably BuNo 83992, ex-AAF 44-85571), was on bailment to Cornell Aeronautical Laboratories beginning in August 1946 where it was intended to mount and test a J-30 jet engine. The specially instrumented test engine was destroyed in a ground mishap and the use of this Fortress as a five-engine testbed did not occur.

Former Pratt & Whitney five-engine B-17 testbed displayed near Hartford, Connecticut, April 1973. The extended nose for turboprop installation is visible beyond the B-17's radial engine cowling. (Author)

This B-17G (44-85747) came to Allison as a Navy PB-1 (BuNo 83999) for use in testing the XT-40 turboprop. The Air Force took charge of the airframe subsequently as Allison used it for the T-56 program. The cockpit was not relocated on this testbed. (AFTC/HO)

52 Testbeds, Motherships & Parasites

One B-17G mounted a man-carrying pod on one wingtip, said to be for the purpose of measuring physiological factors on airmen positioned far from an aircraft's normal centerline. The pod later housed radar units tested by General Electric.

The Army Air Forces' concern over pilot physiology in offset cockpits also manifested itself in the form of a special P-38 Lightning. An early P-38 (40-744) was fitted with a second cockpit on its left boom, replacing that engine's turbosupercharger installation. This predated the P-82 Twin Mustang, with two cockpits spaced far from the aircraft's centerline. The various Wright Field fighter flight test pilots took each other up for aggressive maneuvering in the P-38 with the offset perch. People in the pod included WASP Ann

The pod on this B-17's right wingtip originally was used for physiological testing with a passenger far off the centerline of the aircraft. Later, General Electric adapted the pod for radar testing. (Peter M. Bowers Collection)

This left-side view shows the boom-mounted second cockpit on RP-38 40-744 for physiological testing. Nonstandard engine-mounted exhaust stacks are apparent on this Allison; the new cockpit consumed space formerly used for exhaust piping and exit. (AAF)

This Corsair testbed with experimental jet is on centerline for flight evaluation in 1944. This may be the Westinghouse Yankee. The Corsair vertical fin is marked FG-1. (Peter M. Bowers Collection via Barrett Tillman)

This RP-38 (40-744) mounted a second cockpit on the left boom for physiological tests using an off-centerline crew position. This cockpit necessitated deletion of turbos and use of short exhaust stacks. (AAF)

Chapter 3: Bomber Bonus Testbeds

Baumgartner Carl, who recalled years later that none of the test pilots were bothered by the off-center seating.[55]

If General Electric, Curtiss-Wright, and Pratt & Whitney took advantage of available bomber airframes to validate their jet engines in flight, there is evidence that Westinghouse was less expansive in its turbojet research and development efforts. Westinghouse's early wartime Yankee jet engine made its aerial debut attached to a Navy FG-1 Corsair fighter in 1944, in addition to one B-45 later.

Pratt & Whitney Twin Testbeds after the War

Tests of radial engine developments were facilitated at Pratt & Whitney by putting the test engine or propeller installation on one side of a twin-engine C-46 Commando while retaining the Commando's stock R-2800 on the other side. Similarly, an early Northrop YP-61 Black Widow (41-18888; U.S. civil registration N60358) flew for Pratt & Whitney beginning in 1946, with one stock engine and one test powerplant/propeller. In Pratt & Whitney service, the usual black finish of the P-61 was stripped to bare metal.[56] Damaged while taxiing in 1956, this YP-61 was scrapped.

Pratt & Whitney's 1947 B-29 Testbed

Two years after the end of World War II, Pratt & Whitney obtained a B-29 Superfortress bomber (actually a Navy version dubbed P2B) for use as a testbed carrying the J42 centrifugal-flow jet engine for the Navy's Grumman F9F-2 Panther fighter. The J42 rode in the forward bomb bay. An internal 1,100-gallon wing fuel tank was plumbed to serve the test jet only, isolating its fuel from the gasoline of the B-29. The pressurized gunners' compartment in the aft fuselage housed test instrumentation and controls for operating the jet engine. The flight test program for the J42 concentrated on fuel control and air starts, something that the B-29 testbed allowed without endangering an aircraft, such as the F9F, that relied on the J42 for its only power.[57]

Pratt & Whitney next mounted the more powerful J48 engine, with and without afterburner, to wring it out before it powered aircraft such as the F-94C Starfire and later iterations of the F9F. The availability of testbeds like this B-29 accelerated the development timetable for new jet engines. Before the advent of testbeds, new engines might have to wait until the aircraft for which they were intended was built and ready to fly. Now, jets could be tested proactively and with greater safety before the intended airframe was available.

B-50 Pratt & Whitney J57 Testbed

The Boeing B-50 was a re-engined and substantially modified outgrowth of the B-29. Still called a Superfortress, the B-50 filled an important slot in Cold War Strategic Air Command operational structure until the advent of jet bombers en masse.

An early-model B-50A (46-036), still bearing Eighth Air Force markings, was converted to host a new Pratt & Whitney J57 Turbo-Wasp jet engine in a streamlined nacelle that could be suspended beneath the fuselage in flight. At the time, this B-50 was the largest testbed used by Pratt & Whitney. The J57 could power the testbed B-50 with the bomber's four traditional R-4360 engines shut down. A Pratt & Whitney test pilot, Gil Haven, described the strange and

This B-50A (46-036) helped Pratt & Whitney flight test the J57 attached to a retractable belly mount. The J57 could power the testbed B-50 with the bomber's four traditional R-4360 engines shut down. (Craig Kaston Collection)

54 Testbeds, Motherships & Parasites

initially disconcerting quiet and lack of vibration that came over the B-50 when all of its reciprocating engines were shut down in flight, with only the sole J57 to sustain airspeed.

Extensive flight testing of the J57 on the B-50 preceded the use of this jet engine on the Boeing B-52 Stratofortress and helped give the Boeing team confidence in the engine before their bomber's first flight in 1952.

Pratt & Whitney realized the import of the nascent J57, the first axial-flow jet engine in the 10,000-pound-thrust range. The company's previous experience with B-17 and B-29 testbed aircraft convinced them that the J57 needed that treatment. The B-50 carried the J57 in a bomb bay, extending the engine downward for operation in clear air beyond the airframe's boundary layer airflow.[58]

B-45s for Pratt & Whitney

Like General Electric, Pratt & Whitney comprehended the usefulness of former bombers as testbed hosts for new jet engines. In addition to Pratt & Whitney's five-engine B-17 testbed, its B-29 testbed, and its B-50 testbed, the company made use of two B-45s for test airframes. The B-45 can be deceptive in appearance. It is large enough, with a long bomb bay, to accommodate jet engines, such as the J57 and the still larger J75. The four-jet B-45s achieved higher altitudes and faster speeds than the piston-engine testbed aircraft. Initially, one B-45 serviced the J57 as it matured; the second B-45 at Pratt & Whitney hosted the new J75 engine.

C-124 Globemaster

Another Pratt & Whitney testbed for turboprop engines was made out of a giant Douglas C-124C Globemaster transport (52-1069) in the mid-1950s. It hosted the T57 turboprop engine and propeller combination in a tubular nose cowling much smaller than the size of the mammoth C-124's fuselage. The C-124 was seen at Pratt & Whitney's facility on Rentschler Field in East Hartford, Connecticut, the scene of many Pratt & Whitney flight test activities. The Rentschler airport was closed in 1999 after more than six decades of service.

The C-124's nose was modified to take the turboprop engine and nacelle. The Pratt & Whitney T57 was designed to swing a Hamilton Standard Turbo Hydromatic propeller. Its intended airframe, the unbuilt C-132 transport by Douglas, was canceled in 1957, and the T57 turboprop remained largely an unrealized powerplant.

Testbeds Coming of Age

A writer for Pratt & Whitney summed up the rationale for testbed aircraft eloquently in a 1956 article: "Engine testbeds that have other means of power to sustain flight can maximize the useful run time of the test engine. The testbed can climb to altitude and get on test conditions in advance of the need to power up the possibly costly and scarce test engine. The logged run time on the test engine can thusly be more productive than if the test engine had to power itself up to altitude for test conditions."[59]

Testbed aircraft, especially bombers and transports having great range, can use that range to fly away from inappropriate local weather conditions, allowing test engineers to seek the optimum weather for test purposes. Testbeds of this kind can also stay on test conditions while the test engine is exercised, providing more accurate data and saving valuable test time. Moreover, the inherent safety of having redundant power for a testbed can be a lifesaver (literally) if the test engine experiences serious problems in flight.[60]

The use of flying testbeds was seen as a way to shorten the time between early engine testing on the ground and final evaluation in the intended aircraft type. Testers knew that the value of ground runs, although important, was limited. Even ground test facilities that could simulate altitude conditions were considered inferior to actual flight in real-world conditions. Only a flight test at realistic conditions could build the confidence necessary to deploy a new engine type. Wright A. Parkins, Pratt & Whitney's general manager in 1956, said: "The primary function of a flight test is to get a new engine aloft at an early stage in order to discover and correct its operational weaknesses before production begins."[61]

Sperry Gyroscope Test Fleet

The Sperry Gyroscope company was known for instrumentation, autopilots, the famed ball turret, aircraft gunsights, and even wartime bombsight alternatives to the famed Norden device. Sperry operated a fleet of testbed aircraft at its facility at MacArthur Airport near Lake Ronkonkoma in eastern Long Island, New York. By 1948, the Sperry testbeds included several DC-3s, a B-17, a B-29, and others. In 1953, Sperry took delivery of a Sikorsky S-55 helicopter for its research fleet.

B-26 Bicycle

The post-war crop of jet bombers with high, thin wings inspired new ways to mount landing gear. Martin devised tandem landing gear, with single mainwheels widely spaced fore and aft beneath the fuselage. Variations on this theme graced the Martin XB-48 as well as Boeing's XB-47 and B-52. Since wing-mounted jet engines were essentially long functional tubes within the length of their nacelles, main landing gear could no longer be stowed in wing nacelles, as they had been in so many piston-engine airplanes; space in jet nacelles was at a premium. Like a bicycle with outrigger training wheels, Martin's narrow-track tandem main gear required small balancing wheels on the wings. However, because fuselage wheels in tandem carried the brunt of the bomber's weight, the wing outriggers could be small and light enough to stow in engine nacelles or even in the wings.

Working with the Army Air Forces, Martin developed the tandem landing-gear arrangement to answer multiple urgent demands of post-war jet bombers. Because the jet bombers' fuselages could ride close to the ground, fuselage-mounted tandem landing gear afforded shorter and therefore lighter struts. The tandem main gear was far less dependent on being placed near the aircraft's center of gravity than was conventional landing gear. The forward gear could be well ahead and the aft gear well behind the center of gravity, leaving the bomb bay at the center of gravity. This allowed bomb loads to be dropped without causing undue shifts in the aircraft's center of gravity.

A quirky testbed demonstrator arose for such landing gear arrangements, in the form of a Martin TB-26G Marauder (redesignated XB-26H) bomber that the builder revamped as a tandem-wheel testbed. Whereas jet bombers used dual mainwheels in each position, the converted B-26 used single

Martin's XB-26H testbed proved the soundness of bicycle main gear mounted fore and aft in the fuselage of a bomber. The main gear on the testbed was not retractable, but outriggers in each nacelle could be drawn inside and covered with the landing gear doors. External longitudinal bracing on the fuselage of this one-of-a-kind testbed was probably a safety hedge due to the relocation of landing loads from the wings to the fuselage. (Martin via Stan Piet)

This nearly frontal view shows the XB-26H "Middle River Stump Jumper" landing gear testbed's bicycle mainwheels to advantage. Outriggers in place of the original main gear in the nacelles gave the testbed a four-wheel stance. Boeing adapted this style of landing gear for the B-47, with smaller outrigger wheels extending from the two-jet pods on each wing. (Martin via Stan Piet)

The Martin XB-26H (nicknamed "Middle River Stump Jumper") validated the landing gear arrangement for Martin's own XB-48 bomber as well as for the mass-produced Boeing B-47 Stratojet. (Stan Piet Collection)

Martin test pilots deliberately punished the tandem gear in the XB-26H, making harsh yawed landings and abrupt ground-turning maneuvers. The gear held. Martin's Middle River, Maryland, location gave rise to a nickname for the odd Marauder with the bicycle gear. (Martin via Stan Piet)

mainwheels positioned optimally to serve as the aircraft's main gear. Two much smaller outrigger wheels on long struts occupied space in the Marauder's engine nacelles that were once reserved for the main gear.

Marauder with a Jet

In the immediate post-war years Martin testers made use of several aircraft built by the company. A Navy version of the B-26 Marauder, designated JM-1, housed a Westinghouse 19B turbojet in the aft fuselage with the inlets mounted low on the side of the body in the vicinity of the Marauder's waist gun openings.

The ducted jet engine rode in the fuselage beneath the horizontal stabilizer

Chapter 3: Bomber Bonus Testbeds   57

and elevator, its exhaust exiting the JM-1 where a pair of .50-caliber tail guns had originally been standard. In the Marauder's bomb bay, a droppable fuel tank carried the jet's gasoline (not kerosene) for the test involving a jet burning avgas. Octane ratings from 60 to 100 were tried.

The Navy wanted the flight testing of this jet engine, and made the JM-1 available to Martin. Tests included observations on starting and acceleration at different altitudes and speeds; performance quantification; windmilling characteristics; and specific parameters in climbs and dives. The

The Westinghouse 19B turbojet was hosted by this Martin JM-1 Marauder (BuNo 66599) testbed for research conducted by Martin on behalf of the U.S. Navy in June 1946. Inlets on both sides of the fuselage used former waist gun openings. (Stan Piet)

This U.S. Navy JM-1 (BuNo 66599) Marauder in June 1946 shows a jet inlet mounted in the former waist gun window. The jet exhaust from the Westinghouse 19B undergoing flight testing passed out the rear of the fuselage. (Martin via Stan Piet)

The Martin Marauder bomber made a neat and trim testbed for a Westinghouse 19B turbojet engine mounted in the aft fuselage for Navy tests in 1946. (Stan Piet)

Testbeds, Motherships & Parasites

Martin's venerable A-30 testbed ran its engines in the summer of 1946. The Dorsal fixture was part of the company's research efforts to probe the mysteries of transonic flight by diving the A-30 and achieving accelerated airflow over the fixture. (Martin via Stan Piet)

jet engine rested in a moving cradle that employed thrust meters to measure actual thrust produced by the jet.

Three testers rode in the waist compartment of the modified Marauder. They controlled jet engine operations and had access to firefighting equipment, should it become necessary. The jet was tested at altitudes up to 25,000 feet in climbs, dives, and with the jet's throttle wide open.

The Westinghouse 19B engine produced about 1,400 pounds of thrust at sea level. The later J30 version produced 1,600 pounds of thrust, and two J30s propelled the McDonnell FH-1 fighter. The first Westinghouse Model 19 jet engine made its airborne debut mounted to a Corsair fighter in 1944.

In post–World War II France, a version of the Atar turbojet hitched a ride on a French Air Force B-26G (43-34584). The genesis of the Atar comes from the World War II German BMW 018 turbojet, improved upon by German engineers in French employ.

Martin A-30 Knocks on Transonic Door

The unlikeliest Martin testbed aircraft was a tailwheel-equipped A-30 Baltimore bomber that the company kept on hand into 1947, calling it a transonic testbed. With a top speed in excess of 300 mph, the rugged A-30 could be dived at high speeds to acquire data on test fixtures as Martin joined the general aeronautical quest for flight information to help tame the transonic and eventually supersonic environments.

Classic wind tunnel data of that era for the transonic speed region could be fraught with errors due to the artificial constraints of the tunnel walls. Martin's Aerodynamics Group decided to employ the A-30 bomber as a free-air data-gathering machine.

Pilot Ray Nessly took the silver A-30 aloft to 30,000 feet, diving it 73 times. According to Martin news stories, the

Pilot Ray Nessly (wearing helmet and goggles) and flight test engineers Horace Davis (left) and Tom Woersching (atop fuselage) with the Martin A-30 and its transonic test fixture, 22 April 1946. (Martin via Stan Piet)

aircraft reached 547 mph in a dive. This translated to .72 Mach, or .72 times the speed of sound. The acceleration experienced by airflow over the top of the fuselage gave a localized airflow speed of .84 Mach. At the location where that accelerated airflow occurred, Martin mounted a test fixture with sensors to capture data.[62]

Ten or more different airfoil profiles were milled out of solid duralumin alloy and could be mounted vertically on the dorsal surface of the A-30's forward fuselage just aft of the cockpit. Thirty-three small holes drilled into the solid wing shape accommodated pressure readings. Tubes led from the holes to an illuminated instrument board in the fuselage. Behind the wing section, mounted horizontally, a pressure rake with about 80 small probes measured air pressure. Three 35-mm motion picture cameras recorded the instrument board in an era before telemetered data came from test aircraft. The films showed instrument readings that were plotted in a darkened room with a projector, to give usable data.

Several factors influenced Martin's choice of the A-30 for the high-speed data runs. Company familiarity with the design didn't hurt, and engineers wanted a twin-engine airframe with no prop wash to affect airflow over the fuselage where test shapes and sensors would be mounted. The Baltimore was characterized as robust in construction. Because the A-30 served with Allied air forces and not the United States, Martin was able to acquire one from England through the auspices of the U.S. Navy, the partner in the transonic data-gathering operation.

Allied Signal/Honeywell 720 Testbed

Allied Signal flew Boeing 720 N720GT to test the TFE731-40 turbofan engine in the 1980s. The TFE731-40

Martin flight test engineers Horace Davis (behind) and Tom Woersching (front) adjusted Bell & Howell motion picture cameras that recorded data during high-speed testing with the company A-30 testbed in this April 1946 photo. (Martin via Stan Piet)

60   Testbeds, Motherships & Parasites

Pratt & Whitney used B-52E (56-0636) to mount a large turbofan engine in place of two regular J57s, providing flight test capability for the new turbofan while retaining redundant power from the other six J57s of the B-52E. (AFTC/HO)

traces its lineage to Garrett AiResearch (later Allied Signal and finally Honeywell) as N720H by 2000. Honeywell tested the AS907 engine. The 720 was a former TWA and Northwest airliner before serving overseas companies. As a testbed, it featured a fifth engine pylon on the right side of the fuselage. This 720 testbed was dismantled at Sky Harbor Airport in Phoenix, Arizona, in 2008.

Honeywell 757 Testbed

The follow-on to the Honeywell 720 is a Boeing 757 (N757HW) that also features a special pylon on the right side of the forward fuselage for engine tests. This Honeywell research platform sometimes performs other types of developmental tests with the spare third engine pylon empty. Honeywell acquired the 757 in 2005 and began using it with a third engine in 2008. Engine testing included developments of the TFE731 series and the Honeywell HTF7000 (redesignated from the earlier AS907). The HTF7000 is earmarked for business jets such as the Bombardier Challenger, some Gulfstream models, and others.

B-52 Testbeds

High-bypass jet engines with huge diameters found a home under the right wing of at least two B-52 Stratofor-

tresses. An E-model, 56-0636, served Pratt & Whitney in the 1970s. Replacing the normal complement of two J57 engines on the inboard pylon, the test engine enjoyed the requisite redundancy of the B-52's remaining six original engines. Similarly, General Electric flew B-52E (57-0119) in the 1960s with a gigantic TF39 turbofan in the same position, in preparation for that powerplant's use on the huge C-5 Galaxy transport.

What becomes of old testbed aircraft? In the case of the modified GE B-52E (70119), it lingered at Edwards Air Force Base. Eventually, stripped of useful parts and still bearing vestiges of a test engine pylon on its right wing, 70119 was towed into the desert on the Edwards range. When subsequent nuclear treaties quantified the number of available B-52 bombers, it was deemed to be necessary to destroy 70119 to show that it was not capable of being a nuclear bomber again. As this is written, 70119's blasted carcass rests in the Mojave Desert sun in the company of another derelict Stratofortress, RB-52B 53-0379.

When General Electric sought a testbed for the large and heavy TF39, it winnowed the field down to two possibilities. A C-141 Starlifter or a B-52 Stratofortress were the only aircraft with appropriate weight lifting and high-wing ground clearance to make a viable testbed. Airframe availability favored the B-52. This was the mid-1960s (testing began in 1967) and C-141A Starlifters were brand-new jet airlifters needed for global commitments, including ongoing airlifts to and from Southeast Asia. GE engineers made a test plan that called for speeds as high as .9 Mach and altitudes of 50,000 feet. That also ruled out the Starlifter.

Meanwhile, some older B-52 models were less in demand. A B-52E with 5,000 hours on its airframe was assigned to General Electric to become its TF39 testbed.[63]

The GE test program kept the weight of the B-52 within bounds to permit safe flight on the remaining six Pratt & Whitney J57 engines, should it become necessary to shut down the test TF39 in flight. In 1967, a GE engineering test pilot noted with understandable pride, "I am happy to say

Chapter 3: Bomber Bonus Testbeds   61

The Pratt & Whitney B-52E testbed carried a JT9D turbofan engine on the inboard pylon of the right wing. This arrangement first flew in June 1968. The JT9D had been a contender for powering the C-5A, but that selection went to GE's TF39. The JT9D subsequently powered the Boeing 747 and other wide-body jetliners. (Pratt & Whitney)

General Electric testbed B-52E (57-0119) lands at Edwards AFB with an outsized TF39 engine intended for the C-5 mounted on its right inboard pylon. Modifications to make this a viable testbed eliminated it from USAF inventory, and it was parked in the desert at Edwards. (AFTC/HO)

By 1996, the GE B-52E testbed for the C-5 engines was broken in pieces on the Edwards AFB range, in keeping with a treaty on nuclear-capable aircraft. (AFTC/HO)

The GE TF39 fanjet dwarfs normal engines on the testbed B-52 that helped validate it for the C-5 Galaxy. The huge amount of thrust from the TF39 necessitated some adjusted throttle settings on the other engines to keep the big B-52 on proper test conditions. (AFTC/HO)

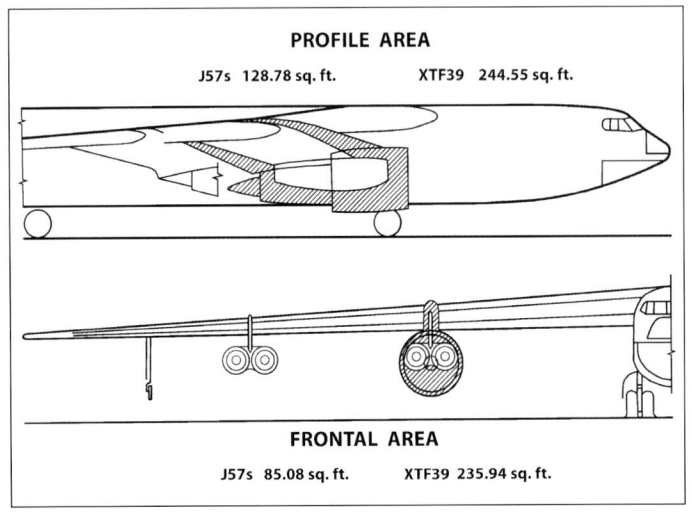

This is a comparison of standard Pratt & Whitney J57 turbojets and the GE XTF39 turbofan test engine on B-52E. (Courtesy of Craig Kaston)

that to date it has not been necessary to rely solely on the brand-X engines to get us home."[64]

The massive TF39 necessitated some interesting flight procedures to accomplish test points under specific flight conditions. GE test pilot Charles R. Anderson explained, "When high power settings are maintained on the TF39, it is necessary to reduce power on engines 7 and 8 (outboard of the test engine on the right wing) to prevent flying large left-hand 360-degree turns."[65]

Flight testing the new and powerful TF39 required extensive engineering to ensure that the testbed aircraft could safely accommodate the large test engine under the most trying circumstances. It was far more than merely bolting a new engine to an old pylon. Wing-skin doublers were added to the upper and lower surfaces of the wing, as was internal wing structure. GE pilot Anderson gave a cerebral explanation of the challenges and the solutions in a 1967 presentation to the Society of Experimental Test Pilots:

"The most demanding phase of the program was the modification to the structure. The design, fabrication of the torque box, fabrication of the internal wing structural components, and strut fairings was accomplished at the General Electric Edwards Flight Test Center and required some 65,000 man-hours. The program was originally based upon the premise that a TF39 nacelle would be attached to the standard B-52/J57 pylon. A more detailed study of the installation in conjunction with loads analysis, established the necessity to completely replace the J57 pylon and to strengthen the wing internally. Basically, we were replacing two engines whose MRT [military thrust] was approximately 21,000 [pounds] and weighed a total of 8,750 pounds with a single engine that developed 41,000 pounds [thrust] and weighed 7,010 pounds.

"Although the test engine weight was less, the basic objective in designing the new pylon was to introduce additional load paths to prevent local wing and pylon overloads created by the increase in frontal and profile areas of the TF39. . . . The original J57 installation contained a frontal area of 85.08 square feet and the TF39 contained 234.9 square feet, or an increase of 277 percent. The J57 profile contained 128.8 square feet and the TF39 profile contained 244.6, or an increase of 189 percent.

"The original Boeing criteria provided for an ultimate side-load factor of 3.0 g. This limit would have been adequate for the new installation; however, since weight was not a factor due to the loss of fuel weight in the wing, this was increased to 4.5 g during the redesign, which produced an even greater margin.

"The requirement to go into the wing can be more fully appreciated by consideration of the following factors. It is possible to visualize greater side-load requirements because of the larger side area exposed to aerodynamic lift and side gusts striking the engine and pylon plus centrifugal force due to airplane lateral roll. Pitching of the aircraft in turbulent air produces greater side moment because the gyro forces of the more massive TF39 rotors exceed the gyro forces of the smaller J57 engines. This moment can conceivably occur in combination with a side gust. Also, increased side stiffness is needed to reduce the possibility of a situation where the gyro forces result in a whirl mode coupled with a wing bending or torsional mode that would produce disastrous results.

"Major considerations in the wing mod/pylon design were to provide necessary strength to accommodate all flight loads with adequate margin and keep changes to wing stiffness at a minimum."[66]

From the description of internal wing modifications plus a special pylon for the GE testbed B-52E, it is easy to see why this older member of the Stratofortress family never returned to the Strategic Air Command bomber fleet.

When the flight test program began, the GE crew had to accommodate the amazing amount of thrust the TF39 produced even at idle. Test pilot Anderson described it:

"The [TF39] engine develops approximately 2,100 pounds of thrust at idle under standard conditions but appears to be developing much more. The influence of idle thrust was one of the factors that would require a closer look-see during the flight phase. During landing flare we could expect approximately 1,600 pounds of idle thrust on the left wing and 2,900 pounds on the right wing. On the first two or three flights the effect was not readily apparent due to strong, gusty surface winds at the time of landing. On subsequent flights, when wind conditions were relatively calm, it became obvious that 4 or 5 degrees of right rudder deflection was required for directional control, and the aircraft would continue to 'float' an additional 1,000 to 1,500 feet down the runway before bleed-off was sufficient for touchdown."[67]

In recent years, General Electric has equipped company-owned 747s to test new jet engines, both large and small, in this time-honored testbed way (see Chapter Four).

B-58 J93 Mount for XB-70, XF-108

The Air Force made ambitious plans for a supersonic bomber and fighter to be powered by the GE J93 turbojet engine. North American Aviation was selected to design and build both the XF-108 Rapier fighter and the XB-70 Valkyrie bomber, but the fighter program was abandoned before an example was constructed.

Growing from GE's successful J79 engine, the larger J93 would propel the XB-70 at Mach 3, 2,000 mph. Plans were made to test the J93 in flight at Mach 2, using a YB-58A Hustler (55-0662) as the testbed aircraft. Central to the B-58 design was the ability to mount a large centerline store beneath the fuselage. This capability lent itself to attaching a custom-made nacelle to the belly of the B-58 housing a J93 turbojet.

This B-58 has the special centerline pod for J93 jet engine testing for XB-70. (USAF)

This B-47E (53-2104) used by the Navy became the unusual seven-engine testbed for GE's testing of the TF34 fanjet. The TF34 attached to the Stratojet's left wing between the B-47's regular J47 engine pylons. The TF34 went on to power the Navy's S-3 Viking and the Air Force's A-10 Thunderbolt II. (General Electric)

TF34 testing included operations on the wing of this EB-47 Stratojet that was serving the Navy for a variety of test and evaluation roles. (General Electric)

This North American AJ-2 Savage (BuNo 130418) was Avco Lycoming's testbed for the YF102 fanjet in the 1970s. The engine, intended for Northrop's A-9, went on to be developed into a powerplant for business jets. The sole remaining Savage later went to the National Naval Aviation Museum in Pensacola, Florida. (Avco Lycoming via U.S. Navy)

Chapter 3: Bomber Bonus Testbeds   67

B-47 Boosts A-10, S-3 Aircraft

Another GE testbed aircraft was B-47E (53-2104), which the company used during development of the TF34 turbofan engine. The TF34's notable successes have been powering the Fairchild Republic A-10 Thunderbolt II and the Lockheed S-3 Viking.

The B-47 testbed eventually became a display in the Pueblo Weisbrod Aircraft Museum in Pueblo, Colorado.

Sole Savage

Avco Lycoming employed a North American Aviation ex-Navy AJ Savage bomber (BuNo 130418; civil registration N68667) as an engine testbed circa 1972–1980, using the Savage's bomb bay to house an extendable jet powerplant, Lycoming's YF102 intended for the Northrop A-9. Follow-on tests by Lycoming proved the merit of the YF102's evolved commercial jet engine, the ALF 502. This testbed aircraft survived to become the only remaining example of an AJ Savage at the National Museum of Naval Aviation in Pensacola, Florida.

Havoc Subs for XF-11

Howard Hughes, known for his wide-ranging taste in airplane acquisitions, used a civilian Douglas A-20G Havoc attack bomber (civil registration NX63148) as a flying testbed for his ambitious XF-11 reconnaissance aircraft. The XF-11 first flew on 7 July 1946, crashing and nearly killing pilot Hughes. Two starkly angular, yet industrially aesthetic, vertical fins and rudders were designed for the XF-11. To gain flight experience with this tail shape, one of the fins and rudders was attached and faired onto the A-20, replacing its traditional rounded vertical fin and rudder.

The AJ-2 Savage Navy bomber used by Avco Lycoming to test the F102 fanjet engine carried cartoon-style nose art, as photographed in 1973. (Photo by Frederick A. Johnsen)

Chapter 4
TRANSPORT TESTBEDS

The huge high bypass GE90 fanjet engine stands out from the rest of the powerplants on GE's first 747 testbed (N747GE) at Mojave in the early 1990s. Fan diameter is greater than 10 feet. As of this writing, the GE90 is the world's largest and most powerful jet engine. (General Electric)

The ample volume of usable space inside transport aircraft has made them viable testbed candidates over the years. As this is written, General Electric routinely operates a huge Boeing 747 jumbo jet airliner as an engine testbed at its Victorville, California, facility located on the former George Air Force Base.

GE Jetliners Test Engines Large and Small

Test instrumentation and test crews have plenty of space in the voluminous 747. General Electric typically mounts a new engine to be tested on one of the inboard pylons in place of one of the 747's regular powerplants. This leaves three-fourths of the 747's original power available as safety and augmentation power for flight tests.

General Electric has employed a variety of jet testbed aircraft at its California sites. A Boeing 707-321 (N707GE; formerly N37681) was initially leased for a certification effort for the CFM56-3 engine. General Electric modified the testbed at Mojave for that task. The test engine was mounted in place of the number-2 engine, inboard on the left wing of the 707. The other three engines remained standard production JT3C jets.

Chapter 4: Transport Testbeds 69

Early in its long career with General Electric, testbed 707-321 carried this blue-and-gold paint scheme and the registration number N37681. The jetliner earned its keep testing iterations of the CFM56 engine, which dwarfed the original powerplants of this 707. Here, the testbed taxies past a distant C-133 stored at Mojave Airport in California. (General Electric)

The 707 remained on GE testbed duty from 1983 until it was scrapped at the Mojave airport storage site and boneyard for airliners in 2005. During this time, the 707 mounted CFM56-3, -5, -5C2, and -5BP1 engines. Tests included in-flight water ingestion. This 707 originally flew with Pan American World Airways, passing through a succession of other operators from the United Kingdom to the Middle East, Africa, and Pakistan before entering its testbed phase.[68]

Primary airframe manufacturers have a vested interest in seeing the engines for their new and evolving aircraft designs developed in a timely manner. Airbus lent the third A300 airliner built to General Electric, enabling

GE's Boeing 707-321 jetliner was a company workhorse for testing several models of the CFM56 turbofan, replacing the number-2 engine on the inboard left-wing position. The 707 served General Electric from 1983 to 2005. GE's jet testbed operation flew from Mojave, California, in the 707 era. (General Electric)

Testbeds, Motherships & Parasites

An Airbus A300B2 jetliner with a test CF6 engine model on its left wing flies in a torrent of water released from a KC-135 tanker over the Mojave Desert to test the engine's performance during water ingestion. The standard version of the CF6 powerplant on the right wing provides a margin of safety, giving the crew confidence to subject the test engine to this drastic treatment. (General Electric)

They're similar but different; two models of CF6 engine power this Airbus A300 used by General Electric as a testbed at its base in Mojave, California. This gave the testbed redundancy in the event that something happened to the test engine in flight. (General Electric)

certification of the CF6-80C2 powerplant. General Electric made the modifications to make this a viable engine testbed at Mojave. For its role as an engine testbed, the A300B2 jetliner (French registration F-BUAD) kept its standard CF6-50C2 engine on the number-1 pylon and the test engine on the other wing.

To certify the 747-400 system, the A300 flew the first CF6 engine with FADEC (Full Authority Digital Engine Control). The Airbus A300 GE testbed also performed water ingestion testing and development as well as certification testing for the GE CF6-80E engine. Before General Electric returned the loaner A300 to Airbus in 1992, this testbed had accumulated about 380 flight hours in GE service.[69]

In the early 1990s, the only practical testbed for the gigantic GE90 engine was a 747. General Electric leased another former Pan American airliner, a Boeing 747-121 (N747GE; ex-N744PA). The GE90 was the choice for the Boeing 777. This time, Lucas Aviation of Santa Barbara, California, modified the jumbo jet for its new duties. With the GE90 riding close to the runway in the 747's number-2 engine position, this testbed made its first flight of that program on 6 December 1993. In the years since 1993, this 747

Chapter 4: Transport Testbeds   71

This five-engine 747 testbed gives plenty of room and plenty of power to test GE's CF34-8, looking diminutive compared with 747's regular fanjets. The CF34-8 series powers aircraft such as the Bombardier Challenger. (General Electric)

testbed has tested engines including the GE90-85B, CFM56-7B, GE90-92B, CF34-8C1, CF34-8C5, GE90-94B, GE90-115B, CF34-10E, GP7200, GEnx-1B, Genx-2B, LEAP-1C, Passport 20, LEAP-1B, and LEAP-1A.

As of this writing, GE 747 N747GE was in long-term storage following its final flight in January 2017, mounting a GEnx-1B engine. This 747 has a long history. It was the 25th 747 built at Boeing's massive Everett, Washington, facility, which was created for that program. Its first revenue-generating flight was in March 1970. Before entering storage, it was the oldest active 747 in the United States.[70]

In 2010, General Electric bought a 747-446 (N747GF; ex-JA8910) after Japan Airlines retired it. General Electric considers this its next-generation testbed. Evergreen Aviation Technologies (EGAT) in Taiwan made modifications to this aircraft for General Electric. General Electric invested in EGAT in 1998. The facility is known worldwide for its maintenance and conversion capabilities. The first GE flight was on 12 January 2015.

As of this writing, testbed 747 N747GF has mounted the LEAP-1A and LEAP-1B jet engines.[71] CFM International produces the LEAP engine family. GE Aviation in the United States and Safran Aircraft Engines (formerly Snecma) in France own CFM equally.

C-82 Jet Thrust Reverser Test

In the 1950s, the NACA found aging U.S. Air Force C-82 Packet transports useful for everything from crash simulations to hosting a jet engine with an experimental thrust reverser. A 1956 NACA report explained how an axial-flow turbojet engine in a large nacelle was mounted to a pylon on the underside of the right wing of a C-82, still in Military Air Transport System (MATS) markings. The jet engine installation included a hemispherical target–style thrust reverser. The reverser was deployed when the C-82 was taxiing as well as standing still. The C-82 did not fly during the thrust reverser research.[72]

NACA Technical Note 3665 said, "A maximum reverse thrust amounting to 58 percent of the forward thrust was obtained at the maximum-engine-speed setting." Unwanted ingestion of hot exhaust air that had been diverted forward was also noted, "During periods of thrust reversal, re-ingestion of the reversed hot gases into the engine inlet constituted an operating problem. In addition to raising the temperature levels throughout the engine, the reverse thrust ratio was reduced by as much as 25 percentage points. Taxi tests indicated that at ground speeds of 62 knots the free-stream velocity was sufficient to disperse the reversed gas flow and prevent it from entering the engine inlet with the engine operating at maximum speed."

A simulated wing surface near the thrust reverser was instrumented to record a heat rise of nearly 400 degrees Fahrenheit when the reverser was deployed. The NACA reported: "The thrust reverser used for these tests consisted of a hemisphere spun from sheet Inconel. The internal diameter of the hemisphere was 24.86 inches, which closely corresponds to 1½ jet nozzle diameters." This diameter was derived by the NACA as a reasonable compromise between thrust-reversal efficacy and the overall size of the stowed reverser.[73]

The reverser was installed in two halves that moved through an arc to impinge, or remain clear of, the jet engine exhaust as required during the testing. No effort was made

This C-82 at Cleveland for NACA tests moved, but did not fly, with a large jet engine nacelle mounted to the right wing for thrust reverser testing. (NACA via Dennis Jenkins)

to create an airworthy thrust reverser; the object was to learn about thrust reverser efficacy and actions. Accurate speed during ground taxi runs was computed by counting the number of full revolutions of one of the C-82's landing wheels over a given period of time.[74]

Testbed Lights the Sky

The Grimes aircraft lighting company decked out a twin-engine Beech Model 18 (C-45H) to test many aircraft lighting configurations in flight. The aircraft was used in the Cold War era and was restored following years of inactivity. The Grimes Flying Lab Foundation maintains the well-illuminated testbed for displays and occasional current lighting tests, advancing the use of LEDs in aviation illumination.

BAC-111

Westinghouse operated a BAC-111 twinjet airliner (N162W) as an avionics testbed. Northrop Grumman subsequently acquired this British twinjet airliner for further use. The BAC-111 is a milepost air transport from an era when British aircraft designs, not consortium aircraft, made their mark in the world. First flown in 1963, the BAC-111 was envisioned as a replacement for other European aircraft such as the Vickers Viscount and Sud Caravelle. Hitting the market ahead of the Douglas DC-9 didn't hurt initial sales, either.

The notional jet engine nacelle and underwing pylon served the purpose for ground tests of a jet engine thrust reverser at the NACA's Cleveland facility. Availability of aging C-82 airframes enabled a number of other tests by the NACA, including destruction. (NASA Glenn Research Center)

Many BAC-111 jetliners flew on airlines in the United States; others flew in countries around the world. It is said the BAC-111 succeeded globally because it was not designed to meet a British requirement, but rather an air transport need. Eventually age and European aircraft noise abatement issues caught up with the BAC-111. Nonetheless, this

A BAC 111 testbed aircraft with the nose of an F-35 was at Joint Base Elmendorf-Richardson, Alaska, on 12 June 2015. (U.S. Air Force/SSgt William Banton)

Chapter 4: Transport Testbeds   73

durable airframe continues to serve Northrop Grumman as this is written.

Occasionally fitted with the unmistakable beaklike nose of the Lockheed Martin F-35, the BAC-111 advanced the Northrop Grumman APG-81 radar used in the F-35. In August 2005, this active electronically scanned array (AESA) radar flew for the first time aboard the BAC-111, racking up hundreds of flight test hours.

Three years later, the F-35 radar suite made its debut flight on another unique testbed, Lockheed Martin's CATBird avionics testbed, which was a highly modified 737 airliner. CATBird is the name derived from the aircraft's function, Cooperative Avionics Testbed.

Dash-80 Testbed

Boeing rightly rocked the airline transport world with the 1954 first flight of its Model 367-80, forever known as Dash-80. The obvious forerunner of the 707 airliner series, Dash-80 had a fuselage cross-section 4 inches smaller than that of commercial 707s. After its original development work was finished, Boeing kept the Dash-80 available for testbed work. It was employed in tests of new flap designs as well as powerplants, at one time flying with three different engine types in its four nacelles. The outboard nacelles hosted JT3 engines; the number-2 pylon carried a J75, and the number-3 pylon mounted a J57.[75]

The Dash-80 also tested a fuselage-mounted fifth engine during Boeing 727 trijet development. Because the Dash-80 had a low-mounted horizontal stabilizer and the intended 727 would have a T-tail, the exhaust for the test engine on the aft fuselage of the Dash-80 used an offset stack to put the jet blast above the horizontal stabilizer. A test with one large Pratt & Whitney JT4 engine on one left wing pylon necessitated skewed throttle movements to accommodate the differing responses of the engines.[76]

Reminiscent of Lockheed's use of an early Constellation, Boeing tried different wing and flap components on the Dash-80. Some of the flap devices were temporary add-ons that lacked the mechanism to deploy and retract, and could only be flown in a low-speed range. Boeing's usual jet chase planes were unable to maintain such slow flight, so the company used aircraft such as a small, slower, Beech 18 for chase activities.

NASA looked at blown flaps, with airflow directed over flap surfaces, as a possible means to improve slow-speed characteristics on air transports as the increasing size and weight of such aircraft put demands on traditional flaps.

A NASA report explained: "The continuing trend toward larger and larger transport aircraft, with the corresponding increases in weight and density, has resulted in reduced response especially in the longitudinal direction. This reduction could result in insufficient precision of control for normal service operation, particularly in the more demanding situations of minimum visibility landings. Because of the possibility of marginal flight-path control in the landing approach of these aircraft, methods for improving flight-path control are being investigated. One method that has shown promise is direct-lift control (DLC)."[77]

In 1969, NASA used the Dash-80, modified by Boeing under NASA contract, to evaluate a direct-lift control flap system. The modifications included fixed flap sections: "For this investigation, the 367-80 was equipped with fixed leading-edge slats on the outboard sections and a fixed Kruger flap over the outer half of the section between the fuselage and the inboard engine." Other flap changes were made as well. Bleed air from the compressors on all four engines supplied air to the boundary layer control system. The tested flaps and augmentation had some benefit, but some limitations were experienced due to a rate limit of the flap actuation system.[78]

The Dash-80 served Boeing well. In the late 1960s, when some Boeing prototypes, including the 737, used an olive green accent color with the traditional yellow fuselage paint, the Dash-80 traded its famed brown-and-yellow color scheme for this olive-and-yellow scheme. It was later returned to its original colors and appears that way today in the National Air and Space Museum annex at Dulles International Airport, Virginia.

Convair 990 Shuttle Stand-In

The NASA Dryden Flight Research Center (now Armstrong Flight Research Center) employed a 1960s-era Convair 990 jetliner to test and evaluate landing gear for the Space Shuttle program in 1993–1994 at Edwards Air Force Base, California, and the Kennedy Space Center in Florida. By mounting a Space Shuttle wheel and tire in a special fuselage cavity in the modified Convair 990, evaluators were able to quantify crosswind performance of the landing gear more precisely than previously.

This led to a proposed raising of the crosswind limitation for Space Shuttle landings from 15 to 20 knots. That extra 5 knots could mean the difference between a landing at the Kennedy Space Center (home of the Shuttles) and a landing at an alternate site, usually, Edwards Air Force Base. The stakes were high and varied for the outcome of the tests with the modified 990 jetliner. An off-Kennedy landing could add an estimated $1 million to the cost of the whole Shuttle mission, as the orbiter had to be specially loaded on the back of a 747 mothership (Shuttle Carrier Aircraft, SCA) and ferried across the United States in fair weather. In the end, no Shuttle missions were landed under the higher crosswind capability.

To evaluate the best circumstances for raising the crosswind landing limits at Kennedy Space Center, the testbed Convair 990 landed on samples of various runway surfaces. This led to the resurfacing of the Kennedy runway to minimize Shuttle tire wear. The Convair replicated Shuttle landing speeds, something ground test devices could not. Once the 990's regular landing gear contacted the surface, the special Shuttle wheel sample was hydraulically lowered from its ventral well. During the test, the actions of Shuttle main and nose tires and wheels were measured along with brakes and nosewheel steering; the 990's stock landing gear remained extended. The project pilot for the 990 Shuttle gear tests was Gordon Fullerton, a NASA pilot and veteran of two Space Shuttle missions.

Following use by NASA, the Convair 990 was retired to display at the main entrance to what is now called the Mojave Air & Space Port in Mojave, California.

For this 1990 test of a Space Shuttle braking parachute, mothership NB-52B Balls-8 became a testbed on the long runway at Edwards Air Force Base. The notch in the right wing, a vestige from X-15 days, is readily visible in this overhead view. (NASA Photo by Dave Howard)

Mother Became a Testbed

The Space Shuttle further benefited when a classic mothership, NB-52B 008, switched roles and became a testbed for a program evaluating a drag parachute for Space Shuttle landings in 1990. The bomber mothership landed on both the paved runway at Edwards Air Force Base as well as the huge flat surface of Rogers Dry Lake adjacent to the paved runway. Test landing speeds ranged from 160 to 230 mph. The instrumented B-52 quantified expected performance characteristics of the drag chute when used with the Space Shuttle, and validated it as a worthwhile device for minimizing wear and tear on the orbiters' wheels and brakes while adding to the Shuttle's margin of safety on landing.

Still earlier, the NB-52B validated the parachutes used for recovery of the solid rocket booster engines sloughed off the Shuttle during launch ascent. Starting in 1977–1978 and again in 1983–1985, Balls-8 was the launch aircraft, releasing test articles representing Space Shuttle solid rocket boosters and their recovery parachutes.

Non-Standard C-141s are No Problem

The production tempo for Lockheed's C-141 Starlifter in the 1960s was such that the first four C-141As built were non-standard airframes as improvements were introduced on later C-141s on the assembly line. The Air Force chose not to place these four early Starlifters in regular airlift operations but kept them as NC-141As for test programs. The letter "N" in

The NC-141A Advanced Radar Testbed navigated the east slopes of the Sierras north of Edwards Air Force Base in the 1990s. Below its unique pointed radome is a smaller black weather radar pod.

Lockheed NC-141A 61-2779 hosted different fighter radars during its career as an Air Force testbed. This Starlifter earned the acronym ARTB, or Advanced Radar Testbed. (Lockheed Martin)

the designation indicates an aircraft in permanent test status. Long after all operational USAF C-141s had been stretched and modified to become C-141Bs, the four early birds remained the only short-bodied A-models in the Air Force.

The Air Force's Advanced Radar Testbed (ARTB) was NC-141A 61-2779. It was fitted with a pointed, elongated fighter-style nose in which any of several modern radars could be tested in an electronic countermeasures (ECM) environment. The spacious Starlifter fuselage offered roomy, if spartan, accommodations for test engineers and their equipment. Radar could be mounted in the long nose when greatest replication of the intended aircraft host was desired, or it could be bench-mounted for access in flight, depending on the needs of the test sortie. Lockheed Aeronautical Systems Company (LASC) made the conversion for the Air Force's Aeronautical Systems Division. The ARTB initially operated as part of the 4950th Test Wing at Wright-Patterson Air Force Base, Ohio. In the early 1990s, all four NC-141As and their mission transferred to the Air Force Flight Test Center at Edwards Air Force Base, California.[79]

Before the end of the decade, all four NC-141s were retired. The ARTB was retained by the Air Force Flight Test Museum at Edwards Air Force Base; the number-1 Starlifter, 61-2775, was ferried to the Air Mobility Command Museum at Dover, Delaware. The other two NC-141As were sent to Davis-Monthan Air Force Base, Arizona, where they were scrapped.

Winging It with a 757

Boeing employed the first 757 airliner (civil registration N757A) carrying a winglike structure mounted atop the forward fuselage to allow airborne testing of F-22 integrated avionics systems under development by Boeing for the Lockheed Martin stealthy fighter. A simulated F-22 nose and cockpit were installed in the 757 to facilitate the work.

Using the 757 for these tests, starting in 1999, was seen as a way to reduce risk and offload some flight test hours that otherwise would have to be borne by the small F-22 test fleet. It allowed the complex avionics suite to be tested in the air before it was installed in an F-22. The winglike structure had the same sweep as the wing of an F-22 and allowed avionics components to be placed and oriented as they would be in an actual F-22.

The Beast of Burbank

One of the earliest examples of a re-purposed heavy transport is "The Beast of Burbank," the original Lockheed XC-69 that later was cut into three sections and spliced with inserts to stretch it as the prototype Super Constellation. This first Constellation originally flew under civilian registration while wearing AAF camouflage olive-and-gray paint and U.S. national insignia. It paved the way for additional wartime orders and construction of small numbers of C-69s.

When the Constellation and the B-29 Superfortress were faced with problems in their R-3350 engine systems, the XC-69 was earmarked for use as a testbed for R-2800 engines instead. But the R-3350s improved and ended the need for the R-2800 Constellation test. The XC-69 rested under tarps from 1944 until war's end a year later. Some contemporary accounts suggest that it retained the test R-2800 engines at that time.

Howard Hughes and Hughes Aircraft had a long-standing interest in the Constellation airliner program. Hughes bought the XC-69. After ferrying it from Burbank to the Hughes airfield, the XC-69 made only one more flight in about a year.

Part of the vast fleet of aircraft Howard Hughes owned at any given time, the XC-69 was purchased by Lockheed from Hughes for use in prototyping the stretched Super Constellation Model 1049 in 1950. As such, this testbed Constellation was known as 1961S (for stretched) at Lockheed. Engineers figured to save substantial money by stretching an existing Constellation to validate the Super Constellation design, and as it turned out, only the XC-69 was available. The other early C-69s and Constellations were earning their keep.

The first Constellation was about to become the first Super Constellation as well. The conversion took less than five months to execute. It first flew as a stretched Super Connie on Friday, 13 October 1950; no one could accuse the Lockheed staff of being superstitious.

For its duties as the 1049 testbed, this Connie flew with bare-minimum furnishings and no pressurization. Plumbing allowed water to be pumped back and forth in fuselage tanks to change the center of gravity for test purposes. Along the way, to better mimic the aerodynamic properties of a full-on Super Constellation, the canopy of the XC-69 was built up and boxed with aluminum and wood to approximate the raised forehead of the Super Constellation, mandated by larger windscreen glazing. It flew with many engine versions and sub-models, and by 1953, the Beast of Burbank was known to be operating on two variants of the R-3350 engine on the outboard nacelles and two R-2800s inboard. It tested several types of Curtiss and Hamilton Standard propellers. The Beast became the first large transport to fly with tip tanks.

In addition, it carried shapes representing the huge dorsal and belly radomes designed for Navy WV-2 and Air Force RC-121 early-warning picket versions of the Constellation. The shapes, as with the installation of tip tanks, provided real-world full-scale validation of these design features, and afforded testers the ability to map the aerodynamic qualities of these add-ons.

The Beast of Burbank finally outlived its usefulness as a testbed in 1958. Sold to California Airmotive, this first Connie was dismantled for parts.

Chapter 5
CANADIAN TESTBEDS

Before its first flight on 30 May 1961, the PT6 testbed Beech 18 warmed up all three engines. The testbed later toured the United States, helping Pratt & Whitney Canada tout the virtues of the remarkable PT6 powerplant. (Toronto Aerospace Museum DHC Photo Collection via Kenneth Swartz)

The United States' neighbor to the north built many significant and successful aircraft, from bush and floatplanes used in rugged arctic regions, to sleek supersonic interceptors, and even the first jet airliner to fly on the North American continent. Yet, when flight test duties called, myriad U.S.-built aircraft of all types were recruited for service as aerial testbeds to aid in the development of Canadian aviation.

Stratojet North of the Border

When Canada needed a suitable airframe to safely flight test its new Orenda Iroquois jet engine, a B-47B Stratojet was lent to Canada and temporarily painted in Royal Canadian Air Force colors in 1956. The Iroquois mounted to the aft fuselage in a neat cowling protruding from the right side of the fuselage. The Stratojet, 51-2059, carried the tail

The Canadian Stratojet on loan was this B-47B (51-2059) that carried the alphanumeric serial X 059 on its Day-glow vertical fin while testing an Orenda Iroquois turbojet mounted to the right aft fuselage under the horizontal stabilizer. Canadair modified the bomber for this duty and called it a CL-52. (Canadair via Tony Chong Collection)

number X 059 in Canadian use. Canadair modified the B-47 testbed and redesignated the bomber as the CL-52. It was a curious testbed, with the Iroquois mounted far aft and offset to the right enough so that it was said to have caused asymmetrical thrust.

Earlier, Orenda (Avro Canada) created Canada's first flying testbeds for turbine engines by replacing the outboard two Merlin piston engines on Lancaster bombers with neatly faired turbojet nacelles flush mounted to the undersurface of the wing. One Lancaster (FM205) tested Avro Chinook jet engines in 1951; it was scrapped in 1956. A second Lancaster (FM209) mounted two Orenda engines. More than 500 hours were logged with this Orenda testbed. Testing stopped in 1954; two years later the Orenda Lancaster was consumed in a hangar fire. Orenda was the inheritor of the Avro Gas Turbine Division, which grew out of Turbo Research.[80]

Canadian Pratt & Whitney Testbeds

Before Pratt & Whitney Canada delivered the first production PT6 turboprop engine in

This leased Beech 18 with a Pratt & Whitney Canada PT6 turbine in the nose is on an early flight, circa 1961. A decade later, Pratt & Whitney bought the aircraft outright. (Toronto Aerospace Museum DHC Collection via Kenneth Swartz)

Chapter 5: Canadian Testbeds   79

1963, this ubiquitous powerplant was test flown by mounting it on the nose of a silver-and-white Royal Canadian Air Force (RCAF) radial-engine Beech C-45 (Canadian HB109), making a trimotor of stark contrasts in motive power. The Canadian testbed started life as an AAF C-45B (43-35478). It served the RCAF No. 32 Operational Training Unit in British Columbia during the war. By June 1953, this Beech flew with the Central Experimental and Proving Establishment's Suffield Detachment.

Still bearing RCAF markings, this C-45 was leased to Pratt & Whitney of Canada on 12 July 1960. De Havilland made modifications. Its first flight as a trimotor with a PT6A engine in the nose was on 30 May 1961. Twenty-three pounds of ballast was taken on board to ensure an appropriate center of gravity.

With the third engine being the inspired PT6 turbine, the trimotor Beech attained test altitudes as high as 26,000 feet. The service ceiling of a stock C-45 was more than a mile lower at 20,000 feet. A Pratt & Whitney test pilot reported the trimotor Beech to have some longitudinal stability issues at high altitudes and high power settings, but the improvised trimotor dispatched its test functions well.

Inflight propeller reversing tests with the PT6 in 1963 produced buffeting of the broad Twin Beech elevator. A quick and prudent decision was made to stop application of reverse power to stop the buffeting. The potential for flutter was a concern.[81]

The trimotor Pratt & Whitney Beech toured the United States in the 1960s to show off the PT6 to American companies designing counterinsurgency (COIN) aircraft that might use the PT6 for power.

Pratt & Whitney Canada bought the aircraft at the beginning of June 1971, registering it as CF-ZWY-X. The Beech logged 1,068 flight hours with the turbine installed. By July 1981, the PT6A had been removed from the nose and the Beech airframe became an instructional tool at the École Nationale D'aéronautique in Québec.

Before Pratt & Whitney Canada decided on the Beech 18 for the PT6 testbed aircraft, the company considered mounting its promising turboprop in the nose of a Douglas DC-3 because the range and service ceiling of the DC-3 suited the test objectives. The rework of the DC-3's nose, however, was considered more complex than that of the Twin Beech, so the smaller Beech became the testbed.[82]

Mallard on a Mission

The turbine PT6 had potential for converting existing piston-engine amphibians such as the Grumman Goose and larger Mallard. Northern Consolidated Airlines converted one of its Mallards to test the PT6. An important area of exploration was the effect of saltwater spray on the turbine. To provide a margin of safety, the testbed Mallard mounted a PT6 in place of the R-1340 engine on the right wing, keeping the traditional piston engine on the left wing. Testing was done at Victoria, on Vancouver Island in 1964. To the relief of the testers, the PT6 showed resilience to saltwater spray, and it kept running even after ingesting large amounts. Turbine Mallard conversions followed.[83]

Viscount Testbed

Pratt & Whitney Canada secured a four-engine Vickers Viscount turboprop airliner for an engine testbed, mounting turboprop engines in the nose of the British transport. The Viscount (Canadian registration CF-TID) had served

Pratt & Whitney Canada's converted Viscount turboprop airliner favored a nose mounting for turboprop tests of PT6 variants. (W. Ross Lennox Collection via Kenneth Swartz)

Pratt & Whitney Canada's Viscount mounted the PT6A-45 for the de Havilland of Canada Dash-7 four-engine airliner in this slightly kinked nose cowl. (W. Ross Lennox Collection via Kenneth Swartz)

with Air Canada when Pratt & Whitney Canada bought it in 1973. The PT6 turbine engine had evolved into the PT6A-50, which was too large for the venerable Twin Beech to accommodate in its modified nose. The Dash-50 version of the PT6 made its first flight mounted to the Viscount on 10 May 1974.[84]

The PT7, redesignated PW100, made its first flight on the nose of the Viscount in February 1982. Other engines in the PW100 series used the perch on the Viscount to demonstrate their performance before powering aircraft. In service with Pratt & Whitney Canada until the fall of 1988, the Viscount logged about 1,250 hours in more than 535 flights. To sustain the Viscount testbed, Pratt & Whitney Canada bought a second Viscount that was stripped for spares and scrapped.

Pratt & Whitney Canada testbeds such as the Viscount and the Twin Beech validated a family of propjet engines that significantly facilitated growth in the shorter-haul parts of the air transport system around the world, giving manufacturers Beech, Embraer, de Havilland, and others the powerplants to enable construction of new-generation short-haul airliners. The Viscount also used its fifth turbine engine to test propeller systems for aircraft, including the Dash-7, Brasilia, ATP, and Fokker 50, which was a reboot of the famous, and aging, F-27 Friendship.

The Viscount CF-TID had logged more than 27,000 flying hours when Pratt & Whitney Canada acquired it. Although it boosted the flight test capabilities for the company, the Viscount's primary Rolls-Royce Dart turboprop engines were running out of life, and its aging fuselage's calendar life limited its pressurization capabilities. It was becoming too slow and altitude-limited to test ever-more-capable turboprops under development by Pratt & Whitney Canada.[85]

720B Testbed

The aging Viscount had served Pratt & Whitney Canada well, but a more capable replacement was needed. The company bought a former Middle East Airlines Boeing 720B that underwent a two-year overhaul and rebuild in England before flying in October 1986. It gained the Canadian civil registration C-FETB, which some wags said stood for Enormous Testbed; an official history says the FETB in the registration aligns with Flying Experimental Testbed. International Aero Engines, an international consortium of jet engine makers that included Pratt & Whitney's American operations, was developing an engine of 20,000 pounds thrust called the V2500, and Pratt & Whitney Canada was set to be the tester with its renewed Boeing 720 jetliner.[86]

The V2500 occupied the 720's modified number-3 engine pylon on the right wing. A blunt nose extension on the red-and-white 720 accommodated turboprop installations. The spacious cabin of the 720 permitted the installation

Chapter 5: Canadian Testbeds

A Canadian Pratt & Whitney turbine engine spins a six-blade propeller on the nose of the company's Boeing 720 testbed in August 2010. (Photo Copyright Pierre Gillard)

Pratt & Whitney Canada's Boeing 720 testbed sometimes flew with a cap on the nose turboprop mount, using a right-hand cheek location for jet engine testing instead. In this 2001 photo, a PW535 business jet engine is carried. (Photo Copyright Pierre Gillard)

of well-instrumented engineer stations for test programs. These interior modifications were made at Pratt & Whitney Canada starting in the fall of 1986 and lasting into early 1988. On 7 May of that year, the 720 first tested an engine, the pylon-mounted V2500. The C-FETB had another trick; a smaller jet turbine could be mounted on a special fuselage pylon to the right of and behind the cockpit.

This red Pratt & Whitney Canada 720 testbed typically operated out of St. Hubert, Québec. The red 720 is associated with flight testing Pratt & Whitney's JT15D, PW300, PW500, and PW600 turbofans as well as versions of the PT6 and PW100 turboprops, plus the PW2037, which powered the Boeing 757-200. The red testbed was sent to Canada's National Air Force Museum in Trenton, Ontario, in 2012.

Pratt & Whitney usually flew a second 720, painted deep blue with a red tail and wingtips, out of Plattsburgh, New York. It carried American civil registration N720PW. The blue jet had a special test pylon mounted to the left side of the aft fuselage.

The blue Pratt & Whitney 720 testbed was retired in 2008, followed by the red bird in 2010.[87]

CF-100 Carried a Test Engine

Pratt & Whitney Canada's flight test program gained access to an indigenous Avro CF-100 Canuck, a straight-wing twin-engine jet, all-weather interceptor with the distinction of being Canada's only production fighter of home design. Somewhat similar to the Northrop F-89 Scorpion in layout, the Canuck had sufficient space for a belly pylon that mounted a third jet engine for test work.

The testbed started life as RCAF 18760, later renumbered (as were many CF-100s) to become 100760. The JT15D was tested on the Canuck. This jet powerplant, with thrust ratings ranging from 2,200 to 2,900 pounds, was used on business jets, including models of the Cessna Citation as well as the SIAI-Marchetti S.211 single-engine trainer.

It is said that the Canuck 100760 was used as a testbed at several times between the late 1960s and early 1980s. It lacked space for full-up data acquisition equipment, and once the Canadian Air Force retired the rest of the CF-100 fleet, support for this Canuck testbed became more difficult. Following its final use by Pratt & Whitney Canada on 28 June 1982, this jet was put on static display at Canadian Forces Base St. Hubert, Québec.

Beech and Lear Testbeds

Pratt & Whitney Canada obtained the oldest extant Beech 200 Super King Air, serial BB2. The second aircraft assembled by Beech as a Model 200 with a T-tail, this airframe actually started life as a Model 100. It came to Pratt & Whitney Canada in October 1979 to augment the venerable Beech 18 trimotor. Instead of mounting a third engine, the King Air 200 (Canadian registration C-GARO) kept a stock PT6 on its right wing and a test engine on the left. Test installations have sometimes shown the King Air with one three-blade and one four-blade propeller among various asymmetries.[88]

For pure jet testing, the Canadian Pratt & Whitney enterprise bought a prototype Learjet from Gates Learjet in November 1981. Formerly carrying U.S. registration N26GL, the Lear obtained Canadian registration C-GBRW. It carries serial number 36-001. As is often the case when aircraft lines evolve, it had helped prototype the Lear models 26, 35, and 36. The Pratt & Whitney testbed had the Lear Longhorn wing with no tip tanks. It was felt that the presence of tip tanks could aggravate flying qualities if fuel sloshed in them during yawing while probing engine handling. In addition, the Longhorn wing gave this Lear improved high-altitude performance.[89]

The Lear retained a stock Garrett TFE731 jet engine on one pylon, and tested Pratt & Whitney Canada powerplants such as the JT15D on the other. The Lear took its mismatched engines to La Grande Rivière, Québec, and two sites in Manitoba for cold-weather testing of the JT15D. As good as the Lear and Beech were, however, they had inherent limitations as testbeds. The Beech's cabin was considered too small for all the desired instrumentation. Moreover, because they are twin-engine aircraft, any time a flight test called for shutting down the test engine, the remaining engine is altitude-limited. These factors argued in favor of the larger Boeing 720 jetliners that followed.

Two 747SPs

Pratt & Whitney based a pair of Boeing 747SP versions as testbeds in Québec in the 21st Century. One of the 747s replaces a normal powerplant with the test engine; the other 747, operated by Pratt & Whitney, can install a test engine on a special stub wing mounted high on the right side of the forward fuselage of the jumbo jet.

Pratt & Whitney's Canadian 747SP testbed (C-FPAW) took the PW1900G engine aloft on the number-2 pylon for the first time in November 2015. (Photo Copyright Pratt & Whitney)

Chapter 5: Canadian Testbeds   83

Chapter 6
POST-WAR MOTHERSHIPS

The silver X-1A rests in its B-29 mothership. From the nose art, it's easy to see how this carrier aircraft sometimes was referred to as the "stork plane." (AFTC/HO)

In the first years of peace following the tumult of World War II, some novel motherships were used for aerial test work. In a unique mix of aircraft types and branches of service, a U.S. Army Air Forces B-17F Flying Fortress was employed by the U.S. Navy to carry and release a sub-scale flyable F8F Bearcat model to evaluate some aerodynamic issues with the Grumman fighter but without risking the loss of a full-size aircraft or its pilot.

The U.S. Navy PB-1 carried an instrumented model of the F8F Bearcat for drop testing, circa 1947. This aircraft, AAF 42-3521, served the Navy as a PB-1W, BuNo 34106. (Peter M. Bowers Collection)

Navy Fortress and Bearcat

In 1947, the Navy employed a stripped B-17F as a model-dropping mothership. The Fortress used for these drops, although originally designated a B-17F-75-DL, initially looked like a chin-turret–equipped B-17G. A number of late-run Douglas-built F-models carried chin turrets before the designation changed to B-17G. This aircraft, AAF 42-3521, served the Navy as a PB-1W, BuNo 34106.

When used as a mothership, the turrets were removed from 34106 and the aircraft flew in basic natural metal finish with extensive anti-glare paint panels. The Naval Air Experimental Station in Philadelphia employed this Fortress to carry a large scale model of a Grumman F8F fighter aloft for drops as part of transonic flight research, a topic of critical importance in early post-war aeronautics development.

Some accounts say that the Bearcat model was weighted with 500 pounds of lead to accelerate its dive speeds into transonic regions in excess of 600 mph, when the automatic controls then pulled the model out of its dive and deployed a parachute for safe recovery. Telemetered data from the Bearcat model and ground radar helped Navy testers quantify the drop flights.[90]

The maturation of the mothership concept came with peace and the quest for supersonic research with a series of rocket-propelled aircraft that needed a lift to preserve their precious and finite fuel for testing at high altitude.

Bell X-1 Needed Boeing B-29

A B-29 Superfortress (45-21800) bearing World War II glossy-black night camouflage on its undersurfaces was modified to receive the Bell X-1 (initially called XS-1) in its bomb bay for post-war tests. No special paint schemes and no particular hoopla attended this workhorse mothership, except for a logo on the nose depicting a cartoon stork carrying the X-1 in flight.

This mothership earned a special niche in history by carrying the X-1 to launch altitude on 14 October 1947, from which the X-1 became the first aircraft documented to reach, and breach, the speed of sound, attaining Mach 1.06 that day.

The B-29 rode high enough off the pavement to enable the X-1 to be nested in the bomb bay, its wings and horizontal tail below the B-29 fuselage. Installing the X-1 in the B-29 required ingenuity. A steep ramp and loading pit were excavated and lined with concrete at Muroc AAF in California to accommodate this process. The X-1 was first backed down the ramp until what remained above ground level could be surmounted by the B-29. Next, in a ginger towing maneuver the B-29 was rolled over the X-1 in its pit at an arc that ended when the B-29 was exactly lined up to receive the X-1 beneath it.

In practice, the X-1 was carried in the B-29 much like a large weapon, with a shackle and special sway braces to

Chapter 6: Post-War Motherships

The P2B-1S Superfortress mothership Fertile Myrtle *rode portable jacks inside the hangar to clear room for mounting the D-558-II Skyrocket. Desert winds could cause problems with outdoor lifting of the Superfortress. The* Fertile Myrtle *nose art is incomplete here. (NASA)*

steady the X-1 and keep it from rolling from side to side in its shackle. With cutout bomb bay doors, the X-1 and the B-29's bomb bay area were exposed to the outside air. A workable if rustic elevator allowed the X-1 pilot to access the rocket plane in flight high over the Mojave Desert.

The NACA described X-1/B-29 launch parameters: "The Bell XS-1 airplane is carried aloft in the modified bomb bay of a B-29 airplane and launched at altitude. The launching characteristics of the Bell XS-1 are entirely satisfactory when the XS-1 is dropped from the B-29 at about –.3 g normal acceleration. The XS-1 is launched from the B-29 at an indicated airspeed of 260 mph so that launching is accomplished well above the stalling speed of the XS-1, which is over 200 mph when the XS-1 is fully serviced."[91]

General (then Major) Robert Cardenas piloted the B-29 mothership on 14 October 1947 when the X-1 first attained supersonic flight. He said that the modifications to make the bomber suitable as a mothership for the X-1 were made by Bell Aircraft.

"The X-1 was hoisted into place using the B-29 internal bomb hoist," General Cardenas recalled. "I could not lift the nose wheel of the B-29 more than about 6 inches without scraping the X-1, which was not a good idea since it was loaded with propellant fuel. It really tested me when I had to land with it still attached. It had to be a soft, near-three-point landing because of the B-29 oleo struts."

Another Superfortress served the X-1 program. For its research flights, the NACA converted a former Navy variant of the type, then called P2B-1S, (BuNo 84029, NACA 137) to host the swept-wing Douglas D558-II Skyrocket.

The old-school method of towing the B-29 over the top of the X-1 in a pit was superseded by huge hydraulic lifts able to elevate the B-29, and later B-50 motherships, high enough to permit the X-vehicle to be towed under it. This jacking operation was subject to the frequent winds in the Mojave Desert, and could only be accomplished in conditions of general calm. Decades after the historic X-1 supersonic flight of 1947, the old X-1 loading pit was retained at South Base on Edwards Air Force Base as a historical site.

The U.S. Navy obtained four B-29 Superfortresses from the Air Force in the spring of 1947. The Navy versions were redesignated P2B in keeping with that service's nomenclature of the time. The former serial numbers were replaced

With its Superfortress mothership on jacks, the Douglas D-558-II Skyrocket number 144 was photographed in the process of being mounted. A large fuselage cutout in the Superfortress accommodates the tall cruciform tail of the Skyrocket. This is the Skyrocket that achieved Mach 2 flight with pilot Scott Crossfield on 20 November 1953. It later became part of the Smithsonian Institution's National Air and Space Museum Collection. (Photo via Boeing)

The swept-wing swept-tail Douglas D-558-II Skyrocket was the most elegant looking of the first-generation rocket planes tested over the Mojave Desert in the late 1940s and well into the 1950s. Three Skyrockets had different iterations of turbojet, rocket, and combined power systems; especially with the rocket power, research with the Skyrocket was accomplished by releasing it from a modified Boeing P2B version of the B-29 Superfortress in Navy markings. A large aft fuselage cutout in the Superfortress to accommodate the Skyrocket's cruciform tail configuration necessitated external bracing on the bomber's fuselage. (AFFTC/HO)

with consecutive Navy Bureau of Aeronautics numbers that apparently had no relationship to the original serials.

Two Navy P2B-1S Superfortress aircraft were intended as motherships. They carried BuNos 84028 (formerly 45-21789) and 84029 (formerly 45-21787). The other two Navy aircraft were called P2B-2S, with Navy BuNos 84030 (formerly 45-21791) and 84031 (formerly 44-87766).[92]

The two P2B-2S variants may have served as testbeds for Navy anti-submarine warfare equipment. The P2B-1S BuNo 84029 gained fame as the mothership for the Douglas D-558-II Skyrocket.

The mothership is usually not viewed as the element of greatest risk in a research project, but forethought about all aspects of safety averted disaster when a Superfortress developed a mechanical emergency. On 22 March 1956 the NACA P2B-1S Superfortress mothership was aloft with the D-558-II Skyrocket attached when the mothership's number-4 propeller, outboard on the right wing, became uncontrollable. Its speed of rotation could quickly lead to mechanical failure. Skyrocket pilot Jack McKay was already in place in the Skyrocket's cockpit when the propeller emergency began. With pre-planned efficiency, mothership pilot Stan Butchart quickly jettisoned the Skyrocket, leaving McKay to set up for a recovery on Rogers Dry Lake.

The errant propeller on the Superfortress came apart rapidly, slinging a knife-edged blade through the number-3 cowling and into the fuselage of the Superfortress where the Skyrocket had been only moments before.

What could have been a disaster for all aboard the combination aircraft was averted. The P2B landed safely, as did the Skyrocket.

The X-1A Predicament

On 8 August 1955, the shiny silver X-1A research rocket was secure in the bomb bay of the B-29 mothership on the ascent to what was supposed to be a flight test high over the hot Mojave Desert. The X-1A and X-1B represented the second generation of the X-1 series. They were ambitiously intended to reach twice the speed of sound and attain altitudes above 90,000 feet. The X-1A and X-1B had revised fuselages and cockpit silhouettes. The X-1A was longer than

Chapter 6: Post-War Motherships

Loading the Douglas D-558-II Skyrocket into the B-29 could be accomplished with the main gear lifted by in-ground hydraulic hoists; the nose gear rested on the lower nose gear hoist pad. At other times, all three hoists were elevated. The vertical fin of the Skyrocket needed clearance to enter the bomb bay cutout. This photo was taken in 1953. (AFFTC/HO)

A runaway outboard propeller on the P2B mothership carrying the D-558-II Skyrocket almost caused disaster for both aircraft on 22 March 1956. The Skyrocket, with pilot Jack McKay in the cockpit, was quickly jettisoned before the propeller blades flung free of their spindles, knifing through the neighboring engine cowl and into the fuselage where the Skyrocket had been moments before. Both Skyrocket and Superfortress recovered safely, albeit the mothership was scarred from the encounter. (AFTC/HO)

A propeller let loose on the P2B Superfortress mothership, and a knife-edged blade sliced through the fuselage where the Skyrocket had been carried moments before. (AFTC/HO)

A trashed engine cowling and propeller hub tell the story of rapid mechanical failure of the propeller on the number-4 engine of the Superfortress mothership in March 1956. The catastrophic propeller separation happened shortly after the manned Skyrocket was jettisoned in a pre-planned safety move. (AFTC/HO)

the original X-1 to accommodate larger fuel tanks for longer rocket engine burns necessary to achieve faster speeds.

The B-29 mothership (which the Air Force Flight Test Center, AFFTC, called a "stork plane") was at 30,000 feet, and the X-1A was about 70 seconds from launch when an explosion ruptured the liquid-oxygen tank, rapidly spewing all of its contents. Several access panels and the landing gear doors blew off the X-1A, and its main landing gear extended into the slipstream. X-1A pilot Joseph A. Walker was not injured and was able to exit the aircraft and enter the B-29.

The AFFTC described the scene: "A small fire started in the wheel well and was observed by the chase pilot, Major Arthur Murray, to burn for about 30 seconds.

"The fire went out and after the X-1A pilot had climbed from the X-1A into the B-29 stork plane, there did not appear to be any immediate danger of another fire or explosion.

An attempt was made to jettison the alcohol and peroxide from the X-1A, but this was unsuccessful. The main landing gear could not be retracted, and it appeared locked down. With the gear in that position, there was insufficient ground clearance to land with the X-1A under the B-29.

"Approximately 30 minutes after the initial explosion, the X-1A was jettisoned on the bombing range of Edwards Air Force Base and it exploded and was destroyed upon impact."[93]

The decision to jettison the X-1A was made by NACA chief research engineer D. D. Beeler and flight operations chief Joseph Jensel because of the extended landing gear, plus the fact that an unknown amount of liquid propellant remained in the X-1A.

The drama was captured by the AFFTC: "When the decision to drop the X-1A was reached, Stanley Butchart, pilot of the B-29, placed his aircraft in a dive to gain speed and then nosed up sharply. At that instant the X-1A was jettisoned. This procedure was followed because the center of gravity in the

X-1A was unknown and there was a possibility that the X-1A might strike the B-29 if level flight attitude were maintained.

"The X-1A dropped tail first, leveled off, and then went into a spin. A second explosion within the X-1A then occurred.

"The wreckage of the X-1A was assembled by engineers of the National Advisory Committee for Aeronautics to determine the cause of the initial explosion."[94]

In the aftermath of this mishap, accident investigators recommended that a primary means to avoid a repeat would be the replacement of a particular type of leather gaskets, which it was believed had promoted a chemical reaction with liquid oxygen leading to the blast. They also suggested some changes to facilitate X-1 pilot egress in emergencies. These changes included, "better means be provided to enable the crew in the mothership to retrieve the pilot from the research airplane in the event of any emergency after the canopy has been secured by the pilot; emergency release mechanism be altered so no sequence could jam the mechanism and prevent a quick launch of the research airplane from the mothership in the event of an emergency."[95]

It was not recorded at the time whether the latter two changes were made; the leather gaskets were replaced on future research aircraft.

X-1D Met Its Fate

Even before the X-1A's perilous demise in 1955, the X-1D met a similar fate over the desert on 22 August 1951 on what was to have been its first powered flight. A loss of nitrogen pressure caused the mission to be scrubbed; pilot Lt. Col. Frank K. Everest attempted to jettison the propellant, a normal procedure for such a situation, but an explosion and fire ensued. The X-1D, without the pilot, was dropped over the desert.

On 9 November 1951, one of the original three X-1s exploded on the ground while nested in a B-50 mothership, causing the loss of both aircraft.

B-50 as Mothership

The B-50 followed the B-29 into the mothership role for the X-Planes. An obvious development of the B-29, the B-50 had been called the B-29D before its nomenclature was changed.

Quick recognition features of the B-50 include a taller fin and rudder than on a B-29 and distinctly different engine cowlings for the B-50's larger R-4360 engines. Early

In a rather rare combination, the third Bell X-1 (46-064) is carried aloft by a later-model EB-50A rather than the B-29 originally used for that task. The aircraft is shown departing Runway 06 at Edwards South Base for this particular X-1's only test flight on 20 July 1951. Note that the X-1's cockpit hatch is left off to facilitate entry by the pilot while in flight. (AFTC/HO)

A lot of support goes into a flight test mission with an experimental aircraft such as the Bell X-2. Test pilot Iven C. Kincheloe posed with the X-2, backed up by the team and vehicles needed to conduct a mission. The red-tailed Boeing B-50 mothership loomed large over everything else. (AFFTC/HO)

Capt. Fitz Fulton smiles beside the nose of the B-50 mothership on 4 September 1956. Fulton earned a reputation as a skilled mothership pilot from the X-2 era through the Space Shuttle 747 carrier aircraft testing and ferrying programs. (AFTC/HO)

By 1956, loading an X-Plane into the bomb bay of a B-50 was made easier by using huge hydraulic jacks to lift the main gear of the mothership instead of towing it gingerly around and over a pit holding the research aircraft. Such lifting operations had to be done in calm air. (AFTC/HO)

Medium shot of the Bell X-2 rocket research aircraft being loaded into the belly of the B-50 carrier in preparation for a flight at the Air Force Flight Test Center, Edwards Air Force Base, California, August 1956.

Chapter 6: Post-War Motherships

Stout hydraulic lifts raised the main gear of the mothership B-50 to enable the X-2 to roll beneath the bomber for mounting in the modified bomb bay. A dolly for the X-2 permits the rocket plane to be pushed into position, keeping the tug vehicle behind the X-2 and out of the way. (AFFTC/HO)

The second Bell X-2 to fly (46-674), arrives at Edwards Air Force Base mated to its B-50 mothership on April 22, 1952. Note the rocket plane's original X-2 logo and "Baby Bell" stork logo on the B-50. (AFTC/HO)

The X-2 has been mated to the B-50 mothership. Evidence of sway braces to keep the wings level is visible. (AFFTC/HO)

Even as the swept-wing Bell X-2 made good use of the B-50's enhanced capabilities as a mothership, plans were already forming for the next mothership that would carry the X-15. A B-50 was considered for the X-15 program, but the giant B-36 was also discussed. Ultimately, the X-15 mothership task went to a pair of early-model B-52s, along with a new way of mounting them. (AFFTC/HO)

Chapter 6: Post-War Motherships

B-50s used nose glazing such as that on B-29s; starting with the B-50D, fewer webs were used on the Plexiglas and the bombsight panel was reshaped.

B-50 and Supersonic Drone

To test the Northrop/Radioplane XQ-4 supersonic drone in 1956, the Air Force used a B-50 mothership at Holloman Air Force Base, New Mexico. A pylon outboard of the bomber's engines on the left wing carried the XQ-4. It was said to be able to reach 1½ times the speed of sound and exceed 60,000 feet.

Powering the XQ-4 was a single Westinghouse J-81 jet engine derived from the British Rolls-Royce RB.82 Soar expendable powerplant. The Soar led to a couple of British testbed and mothership configurations, including a Gloster Meteor jet fighter testbed demonstrated at the prestigious 1953 Farnborough International Airshow.

The Meteor mounted a Soar engine on each wingtip. The Soar was the intended motive power for the British tri-motored Red Rapier missile, with the engines mounted on the horizontal and vertical tail tips. Unpowered, small scale models of the Red Rapier (Vickers 825) were carried aboard a Royal Air Force Washington (B-29) bomber and air dropped over the Woomera test range before the project was canceled.

Air Force Capt. Iven Kincheloe begins to accelerate the X-2 as it drops from the B-50 high over the Mojave Desert on its 10th powered flight on 3 August 1956. Both X-2 aircraft were destroyed during the program, achieving only 20 flights including early glide sorties. The luxury of color photography for some of those missions in the 1950s is a bonus appreciated by historians. (AFFTC/HO)

Capt. Kincheloe will soon fly the swept-wing Bell X-2 away from the red-marked EB-50D mothership that carried it to altitude on 3 August 1956. Kincheloe attained Mach 2.57 that day, the 10th powered flight for the X-2. The extra plumbing visible behind the X-2 remains fixed to the mothership on launch. (AFFTC/HO)

Chapter 7
MOTHERSHIPS IN THE JET AGE

The X-15's balance on the special B-52 underwing pylon keeps the cockpit forward of the wing, making ejection feasible for the X-15 pilot. The rear of the X-15, with its wedge-shape vertical fin, required accommodation by the B-52, so a piece of the bomber's wing trailing edge was removed. (AFFTC/HO)

As the jet engine revolutionized high-performance flight in the late 1940s and early 1950s, the need for jet-powered motherships to launch ever-more-advanced experimental rocket-powered test aircraft became paramount. Beginning with rocket planes that probed the supersonic realm of Mach 1 and Mach 2, larger and more powerful motherships were soon launching "space planes" that left the Earth's atmosphere, even qualifying their pilots as astronauts in the process.

"Higher, Faster, Farther"

The above mantra in aerospace research had a fundamental rationale much more significant than simply entries in the record books. Higher flight could probe the edge of space. It could make aircraft less susceptible to attack in some scenarios. It could provide amazing views and reconnaissance capabilities. Faster flight compresses the time needed to get somewhere. It eludes potential adversaries.

And, in some research cases, it is the means of generating heat by friction for materials research. Farther flight is the continual goal of engineers; extended flight duration can translate into everything from fuel economy to greater mission performance.

Motherships contributed to all three of these goals. Following the successes of the B-29 and B-50 motherships over Edwards Air Force Base, larger and more capable motherships were needed for the next level of X-Plane testing, the Mach 5 X-15. When notions of the X-15 were still gestating, the possibility of using a B-50 mothership was posited. For better performance and load carrying, modifying a giant B-36 bomber to be the mothership was explored. The huge bomb bay of the B-36 was to be the location for the partially exposed X-15. This arrangement would allow the X-15 pilot to enter and exit the research rocket aircraft during flight.

The NACA prepared a preliminary specification for the research aircraft that became the X-15. This October 1954 document said that the research vehicle's weight and size "shall be such as to permit air launching from a mother airplane, such as the B-50, B-36, or B-52, thus effectively providing a two-stage vehicle."

Interestingly, the U.S. Air Force Air Research and Development Command (ARDC), which had the lead responsibility for development and purchase of the X-15, was more circumspect in early descriptions of the rocket plane's mothership. Five months after the NACA's preliminary specification was written, in March 1955 ARDC merely said, "A B-36, B-50, or an airplane of comparable size will be required for modification to the carrier configuration." The Air Force omitted specific mention of a B-52. To expect a new B-52 Stratofortress strategic bomber, then just going online with the Strategic Air Command, to be taken out of operational service for use as a mothership seemed very unlikely.

A March 1956 X-15 System Development Plan report generated for the Air Force still called for a B-36 mothership to be modified by North American Aviation to allow the X-15 to nest partially revealed from the modified bomb bay of the big bomber. The report already estimated that the X-15 program would require 40,000 gallons of aviation gasoline in Fiscal Year 1959 and that amount again in Fiscal Year 1960 to fuel the B-36; even its four jet engines ran on avgas.[96]

There's a celebrated truism in aircraft development known as the pound-a-day increase. It is said that a complex aircraft gains weight at that rate during its development. Although perhaps not statistically correct, it does characterize many aircraft development programs in which early notional optimism must give way to pragmatic equipment and structures necessities, leading to aircraft weight increases.

B-52 versus B-36 as X-15 Mothership

As the X-15 was taking form, its advocates were increasingly worried about performance compromises that this weight would demand. That became one argument in favor of using the B-52 instead of the older B-36. The B-52 promised perhaps a mile higher launch altitude at faster speed than the B-36 could achieve. The B-36 was also in its final years of service to the Air Force, and X-15 programmers worried about the ability to logistically support such a complex bomber as the B-36 without the fleet capabilities and economy of scale once the rest of the B-36s were retired.

Before a B-36 became the X-15 mothership, two early B-52 Stratofortresses were made available for the task. To a generation of people watching the nightly news, or building the latest plastic model kits, these two B-52s became the iconic motherships of all time. One was B-52A 52-003, nicknamed "The High and The Mighty One"; the other was originally a B-52B, 52-008, generally referred to as Balls-8 in the tradition for aircraft with multiple zeroes in their serial numbers leading to a final digit.

As permanently modified, the motherships were no longer viable bombers. Their nomenclature became NB-52A and NB-52B, with the letter "N" designating permanent special test status. The conversion to mothership was estimated to cost $2 million for each NB-52.

The NB-52A flew 59 X-15 missions and was retired first. The NB-52B flew 140 of the 199 X-15 sorties and then went on to a long career as a mothership for lifting bodies, sub-scale models, missiles, and more, mounted to iterations of its special pylon. The NB-52B logged 1,051 flights, accumulating 2,443.8 flight hours. At the time of its retirement in 2004, it was ironically the oldest B-52 flying as well as the B-52 with the lowest flying-hour total, compared to bombers in the regular Air Force Stratofortress fleet.

The B-52 mothership 008 carried the X-15 aloft with a T-38 chase aircraft in proximity. The use of B-52s as motherships revolutionized the practice, taking research aircraft out of the bomb bay and into the sunshine on a wing pylon. (AFFTC/HO)

North American Aviation, builder of the X-15, made the necessary modifications to the B-52s at Plant 42 in Palmdale, California, not far from Edwards Air Force Base. As often happens with aircraft that become testbeds or motherships, the availability of then-new technology (such as B-52s for X-15 motherships) is due in part to the rapid operational advances made in the B-52 fleet. With first flight in 1952 and first operational use by Strategic Air Command in 1955, the two B-52 motherships were barely seven years old when they began their new duties, but their operational B-52 siblings had rapidly advanced beyond them in efficiencies and equipment that the Air Force required.

The A-model did not see operational service, and only 50 B-models were made before production shifted to later iterations. This left a few A- and B-model Stratofortresses as handy, large airframes for tasks other than nuclear deterrence. Balls-8 was only the 10th B-52 built.

The high-wing B-52 changed course from mothership tradition in a salient way; instead of nesting the X-15 in the bomb bay, the two B-52 motherships were fitted with a sturdy underwing pylon inboard of the engines on the right wing of these Stratofortresses. To accommodate the vertical fin of the X-15, a large angular cutout was needed in the trailing edge of the B-52's right wing. A bug-eye window for observation and camera use was added to the forward fuselage of the B-52. The bomb bay of the NB-52 motherships carried tanks filled with liquid oxygen to top off the tanks in the X-15 before dropping the rocket plane. The pylon

The NB-52B (52-0008) mothership for the X-15, sometimes called The Challenger, appears on the ramp at Edwards AFB in the era of Day-Glo orange paint for high visibility. This famed carrier aircraft flew until 2004, subsequently being placed on display at Edwards Air Force Base, the scene of its long and productive career as a mothership and occasional testbed. (AFTC/HO)

The majesty of a giant eight-engine B-52 mothership taking the X-15 aloft into the stratosphere, making it possible for the black rocket ship to use its onboard propellant to attain previously unheard-of speeds and altitudes, was captured looking up through the canopy of one of several chase planes attending each X-15 launch. (AFTC/HO)

was the conduit for the X-15's needs while mated to the mothership.

This wing-mounted arrangement for the X-15 necessitated another change in the way that X-Plane missions were conducted. No longer could the research pilot enter the test aircraft by climbing into its cockpit from inside the converted bomber. Now, the pilot entered the X-15 as it hung from its pylon on the ramp, and he remained there throughout takeoff and ascent to launch altitude.

The advent of a workable zero-zero (zero speed, zero altitude) ejection seat for the X-15 provided a measure of confidence in this new mounting arrangement. The X-15 cockpit rode far enough ahead of the wing of the B-52 to provide a clear overhead path for the pilot in the event he needed to eject while still attached to the bomber's wing pylon.

A 1965 NASA publication, *X-15 Research Results (SP-60)*, illuminates the unique precedents set by the X-15/B-52 combination aircraft: "Many flight tests were made to integrate the X-15 with the B-52 launch-airplane operation. The air-launch technique had been proven, of course, with previous rocket airplanes. The concept has grown, however, from a simple method for carrying the research aircraft to high initial altitude, to an integral part of the research-aircraft operation. For the X-15, the air-launch operation has become, in effect, the launching of a two-stage aerospace vehicle, utilizing a recoverable first-stage booster capable of launching the second stage at an altitude of 45,000 feet and a speed of 550 mph.

"As with any two-stage vehicle, there are mutual interferences. They have required, among other things, stiffening of the X-15 tail structure to withstand pressure fluctuations from the airflow around the B-52 and from the jet-engine noise.

"Several of the X-15 systems operate from power and supply sources within the B-52 until shortly before launch; namely, breathing oxygen, electrical power, nitrogen gas, and liquid oxygen. A launch crewman in the B-52 controls these supplies; he also monitors and aligns pertinent X-15 instrumentation and electrical equipment. In coordination with the X-15 pilot, he helps make a complete pre-launch check of the latter aircraft's systems. Because this is made in a true flight environment, the procedure has helped importantly to ensure satisfactory flight operations. The mission can be recalled if a malfunction or irregularity occurs prior to second stage launch.

"These check-out procedures are also important to B-52 crew safety, since the explosive potential of the volatile propellants aboard the X-15 is such that the B-52 crew has little protection in its .040-inch-thick aluminum 'blockhouse.'

"The launch is a relatively straightforward free-fall maneuver, but it was the subject of early study and concern. Extensive wind tunnel tests were made to examine X-15 launch motions and develop techniques to ensure clean separation from the B-52."[97]

The rationale and hardware developed for the X-15/B-52 combination was sound. With various modifications to

Both X-15s are saddled up and ready to fly on the same day in this photo from 4 November 1960. NASA tried, unsuccessfully, on four occasions to launch two X-15 missions in a day, but not more than one mission was ever accomplished in a day. On this date, Air Force pilot Robert A. Rushworth made his first X-15 flight, reaching a speed of Mach 1.95 and an altitude of 48,900 feet. (NASA)

In 1980, the sub-scale HiMAT remotely piloted research aircraft was photographed in its pylon adapter under the wing of the NB-52B. The adapter is marked for carrying a quantity of JP-5 fuel, which is generally a Navy blend. (NASA)

accommodate different vehicles, B-52 mothership 008 remained in service from 1959 to December 2004. It flew 159 X-15 missions, 140 of which were launch flights; the others were aborts or captive-carry tests.

From 1966 to 1975, the B-52 008 launched 127 of 144 flights of the wingless lifting body research vehicles.

In 1977–1978 and 1983–1985, Balls-8 released models of the Space Shuttle solid rocket boosters to help develop the parachute recovery system for these components.

In the last half of 1990, the 008 switched from mothership to testbed to provide a vehicle for testing Space Shuttle drag chutes. With landing speeds ranging between 160 and 230 mph, the B-52 deployed the tail-mounted Shuttle chute as loads were recorded for analysis. Landings were made on the hard clay surface of Rogers Dry Lake as well as the paved runway at Edwards Air Force Base.

The NB-52B carried and launched a number of remotely piloted vehicles in the 1970s and 1980s. One was a sub-scale F-15 that enabled spin testing without risking a pilot and a full-size F-15. The B-52 also released the Highly

In 1973, the sub-scale F-15 spin-research remotely piloted vehicle was photographed mounted in a special adapter on the X-15 pylon under the wing of the NB-52 mothership. This unpiloted F-15 model enabled spin testing with no risk to a pilot or an expensive full-scale F-15. (NASA)

100   Testbeds, Motherships & Parasites

Riding in an adapted cradle suspended from the NB-52 wing pylon in 1977, the bright orange vehicle is a Drones for Aerodynamic and Structural Testing (DAST) modified BQM-34 Ryan Firebee. (NASA)

Technicians groom a DAST configuration on the NB-52 pylon in 1979. (NASA)

The wing of the B-52 mothership 008 flexes under load as the X-38 drop vehicle hitches a ride aloft, high above Edwards Air Force Base, on 19 November 1997. (NASA/Jeff Doughty)

The final campaign for the NB-52B was the X-43 hypersonic program in which a Pegasus missile, with the supersonic combustion ramjet X-43 mounted to its nose, was dropped from the B-52 over the Pacific Ocean. The Pegasus accelerated the X-43 before separating; then, the X-43 accelerated further on the power of its ramjet, proving the concept as the two successful X-43 flights beat Mach 6 and Mach 9 in 2004. The final X-43 flight in November 2004 was followed by the NB-52B's retirement to the Air Force Flight Test Museum in December of that year. (NASA)

Chapter 7: Motherships in the Jet Age

One of many tasks performed by the B-52 Balls-8 was the carriage and release of iterations of the proposed X-38 crew return vehicle intended for use with the International Space Station. The exploratory X-38 glide models were released by the B-52, sometimes reaching transonic speeds in their autonomously controlled unmanned descents before deploying a drogue parachute and a parawing to aid in gentle landings on the surface of Rogers Dry Lake at Edwards Air Force Base. The NASA X-38 program was canceled in 2002 before an orbital version could be tested.

Both outrigger wheels are off the runway as the flexing wings of the NB-52B mothership Balls-8 roars down the Edwards Air Force Base runway with one of the X-38 development models riding in the modified pylon under the bomber's right wing. A smoky exhaust plume trails the B-52. Because of the cost of converting it to accept the JP-8 fuel almost universally used by the military later in its life, Balls-8 was the last B-52 that required JP-4.

Maneuverable Aircraft Technology (HiMAT) research vehicle as well as supporting Drones for Aerodynamic and Structural Testing (DAST). The B-52 was a viable way to test improvements to the F-111's cockpit capsule parachute system, making several air drops of F-111 cockpits starting in 1979.

According to a NASA fact sheet, starting on 5 April 1990 the first six launches of the commercial Pegasus rocket were from the wing pylon of B-52 008. Between 1998 and 2001, Balls-8 carried and released derivative models of the proposed X-38 space return vehicle. This mothership finished its remarkable career with the X-43 hypersonics program between 2001 and 2004.

Lifting the Lifting Bodies

The B-52 008 proved to be adaptable to carry a series of lifting-body research vehicles on flights from Edwards Air Force Base from 1966 to 1975. The lifting-body research developed piloted aerodynamic shapes without wings that were contenders for space re-entry vehicle follow-ons. The need was classic mothership domain: The NB-52 carried the lifting body aloft, often above 40,000 feet, for release within return distance to Rogers Dry Lake at Edwards AFB. Typically using proven XLR-11 rocket engines, lifting bodies achieved supersonic speeds.

The lifting bodies were conceived to explore specific flight envelopes. Lifting body pilot Maj. Jerauld Gentry described the need: "In the case of the lifting bodies, the M-2/F-2, the HL-10, and the X-24A, the specific flight test objectives were: first, to demonstrate the feasibility of unpowered horizontal landings; second, to explore the transonic and low supersonic speed regimes; and third, to demonstrate an unpowered approach from a representative terminal reentry flight condition [Mach 1.8 at 90,000 feet]."[98]

The aerodynamic requirements for the shape of the lifting bodies argued against employing jet engines, as Major Gentry explained, "Using a jet engine to power the lifting bodies was not given serious consideration. There was no jet engine capable of developing the required thrust for takeoff and acceleration to supersonic speeds, and at the same time small enough to fit into the vehicle without seriously compromising the desired vehicle shape and aerodynamic characteristics."[99]

So the lifting bodies were designed to be launched by the NB-52 motherships, which were proven carrier aircraft. As the lifting-body program progressed, the basic X-15 pylon required adaptors to accommodate varying lifting-body geometries. As Major Gentry described it, "These vehicles were designed for launch from a B-52 mothership at an altitude of 45,000 feet and a Mach number of .8, much the same conditions as the X-15 research vehicle."

Captive-carry sorties under the wing of the B-52 were valuable preparatory efforts, he noted: "One advantage of the air-launched research vehicle is that during first flight preparation the vehicle can be taken on captive flights to

The only way into the M2-F2 on the pylon of the B-52 mothership was to climb aboard before takeoff. Here, NASA research pilot Milt Thompson smiles as he prepares to enter the M2-F2 on 28 February 1966, assisted by Jay L. King, Joseph D. Huxman, and Orion D. Billeter. This flight was a manned captive-carry mission for the M2-F2; Thompson took off and landed with the lifting body firmly attached to the mothership. (NASA)

Chapter 7: Motherships in the Jet Age

As the M2-F3 drops away from the NB-52 mothership, the modified appliance beneath the main pylon, required to carry the triple-tailed M2-F3, is evident. (NASA)

The mothership Balls-3 (NB-52A 52-003, sometimes shown as 52-0003) begins cycling its landing gear upon takeoff from Edwards Air Force Base with the M2-F2 lifting body on the pylon in this 1966 photo. The twin-tailed M2-F2 used a different pylon extension than did the later M2-F3 triple-tail variant. (NASA)

Blunt-nosed X-24A streamed wisps of dust behind as it rolled out on Rogers Dry Lake with a NASA F-104 chase aircraft in close attendance. Test vehicles such as the X-24A were able to accomplish their missions by hitching a ride aboard the underwing pylon of a B-52. The X-24A was designed to explore the low-speed end of the lifting-body performance envelope. (Gene Furnish Collection via AFFTC/HO)

In September 1975, NASA research pilot Bill Dana paused with the X-24B following its final powered flight. The X-24B was essentially the former blunt-nosed X-24A with a sleeved, elongated nose section. (NASA)

verify systems operations at the planned environmental conditions.

"On manned captive tests all systems were given a thorough checkout. A rapid descent from launch altitude of the B-52 with the lifting body in captive position was used to simulate an actual flight; and the ram air turbine was deployed, as was the landing gear. These tests uncovered additional problems to be solved."[100]

Meticulous planning and double-checking attends flight test programs at Edwards Air Force Base. For the NB-52 mothership, this was more than simply hanging a lifting body from the bomber's special underwing pylon. A document from the X-24A lifting-body flight test program mentioned the checks that the B-52 went through: "Prior to the first captive flight with the fully serviced vehicle, the natural frequencies of the NB-52/pylon/X-24A combination were determined by ground tests to be satisfactory [3.2 Hertz in pitch and 3.0 Hertz in roll]. Vehicle/pylon motion was studied during a high-speed B-52 taxi test."

From a close-in chase-plane position, the X-15 and NB-52A mothership combination dominates the view. The upper portion of the X-15's vertical fin is obscured by the cutout in the B-52 wing. The special pylon attached to the wing of the B-52 by four bolts. (USAF via Gene Furnish Collection)

B-52 Pylon Life

NASA put its computational and engineering brainpower to work in the 1980s to study the venerable X-15 pylon on the NB-52B in an effort to understand its fatigue life. When North American Aviation modified the B-52B, the special pylon for the X-15 was attached to the wing with bolts at four points. The strong steel bolts were not considered fatigue points. But the hooks for holding a drop body were subject to fatigue, as were sway braces and drag pins.[101]

A NASA study published in 1985 looked at the pylon and its components. The study was triggered by a failure of rear hooks while moving on the ground with a model Space Shuttle solid rocket booster intended for parachute tests. A NASA technical report gives insight into the pylon's history and construction: "The NASA B-52-008 carrier aircraft has been used to carry various types of test vehicles for high-altitude drop tests. Typical test vehicles carried by the B-52 have been the X-15 (35,250 pounds without drop tanks, 51,600 pounds with drop tanks), HL-10 lifting body (15,380 pounds), HiMAT (3,528 pounds plus 4,000-pound adapter), DAST (2,500 pounds with 4,000-pound adapter), and solid rocket booster and drop test vehicle (SRB/DTV, 49,000 pounds).

"The test vehicle, or adapter, is attached to the B-52 pylon through one front hook and two rear hooks. There are two sets of rear hooks; one set is for carrying longer test vehicles such as the X-15 and SRB/DTV; the other set is for carrying shorter test vehicles such as drones for aerostructural testing [DAST]. The previously established limits for vertical loads on the front and rear hooks are 37,700 pounds and 57,600 pounds, respectively.

Chapter 7: Motherships in the Jet Age 105

On 30 March 2001, the NB-52B mothership Balls-8 was photographed in flight with no drop vehicle visible on the pylon. NASA had kept this B-52 available and airworthy for many years; the desert sun has taken a toll on its paint and markings. By this time, it was by far the oldest B-52 still flying. Rugged and reliable, Balls-8 served until late 2004. (NASA Photo by Jim Ross)

"The SRB/DTV is a scale model of the SRB and is used for testing the performance of the SRB main parachute system. The DTV is 611 inches long and weighs about 48,267 pounds. The SRB/DTV weight will induce static vertical loads of 13,908 pounds at the front hook and 17,180 pounds at each rear hook.

"Although the static hook loads were well below the limit loads, the two rear hooks failed during towing of the B-52 carrying SRB/DTV on a relatively smooth taxiway [very small dynamic loading], after cancellation of the first of the new series of SRB/DTV drop tests because of unfavorable weather conditions.

"The initial failure occurred in the outboard rear hook, resulting in an increased load on the inboard rear hook and causing it to fail as well. The microscopic observations of the fracture surfaces of the two failed rear hooks revealed that microsurface cracks existed at the rounded corners [or hook notches] of both hooks. These surface cracks became unstable and propagated because the stress levels in the vicinity of the crack sites exceeded the critical values associated with the crack sizes."[102]

A 1989 NASA mathematical study of the B-52 pylon's hardware found that "For the front and the two rear hooks the crack growth rate is most rapid during the initial stage of taxiing and the takeoff run and becomes very slow during cruising because of relatively low-amplitude stress cyclings." Other stress points identified in the pylon that were affected by drag had low crack growth in taxiing but higher growth during takeoff and cruise due to increased drag.[103]

This glimpse into the world of mothership design makes it clear that the success of the NB-52s was predicated on

The ill-fated B-52H received at the NASA Dryden Flight Research Center was photographed on 16 April 2002. It was returned to the Air Force for use as an instructional airframe without flying a mothership mission. (NASA)

calculated engineering and inspection; it was far more than merely bolting something onto the B-52.

B-52H Benched

NASA had intentions of keeping a B-52 mothership capability available at Edwards Air Force Base after the retirement of the aging NB-52B 008. Accordingly, in April 2002 at ceremonies in Wichita, Kansas, a newly painted white NASA B-52H mothership (61-0025) was publicly shown during events marking the 50th anniversary of the B-52 Stratofortress. The white B-52H subsequently flew to NASA's Dryden Flight Research Center on Edwards Air Force Base, California, for further modification.

A lack of forecast research projects needing the use of this B-52H led to its return to the Air Force for use as an instructional airframe. The B-52H's previous unit of assignment had been the 5th Bomb Wing.

TriStar Mothership

As large, capable transport jets became generally available, entrepreneurs found ways to capitalize on these vehicles as privately owned motherships. Orbital Sciences Corporation converted a Lockheed TriStar wide-body jet airliner (N140SC) as the mothership for the Pegasus rocket booster. Acquiring the TriStar from storage in Marana, Arizona, in 1992, the aircraft (named Stargazer by Orbital Sciences) first launched a Pegasus rocket on 3 April 1995.

A design feature of the Lockheed TriStar inadvertently made it ideal for this mission, with a structural space between

On 15 June 2004 the B-52H at the NASA Dryden Flight Research Center showed its new mothership pylon, still in primer paint color. The new pylon was designed for anticipated vehicles that would not require the wing cutout of the classic X-15 installation. It was never used. (NASA)

Chapter 7: Motherships in the Jet Age   107

The X-34 program only got off the ground with captive-carry flights beneath the Orbital Sciences Lockheed L-1011 mothership. This image shows the first captive-carry flight at Edwards Air Force Base on 29 June 1999. The X-34 was canceled before actual flight. (NASA)

two longitudinal keel members allowing the vertical fin of the Pegasus to ride inside the airliner, occupying what formerly was a galley.

The Stargazer L-1011 typically launches a Pegasus from 39,000 feet while flying at .82 Mach. In this way, a satellite weighing up to 1,000 pounds can be placed in low earth orbit. The Pegasus launch protocol calls for the Pegasus to free-fall for 5 seconds before ignition of the first stage. The orbit of the satellite payload is typically achieved in 10 minutes after launch.

Orbital Sciences patented this launch system that has so far placed 86 satellites into earth orbit. The three-stage winged Pegasus rocket and its company-owned L-1011 mothership were the world's first privately developed space launch vehicle, following initial Pegasus launches from the pylon of NB-52B mothership Balls-8. Orbital Sciences is fond of calling Stargazer an air-breathing reusable first stage.

The Pegasus, first under a B-52 and then under the L-1011, pioneered the air launching of a rocket from a mothership to put satellites in orbit. Launches are not constrained to traditional sites such as Kennedy Space Center or Vandenberg Air Force Base. The L-1011 can ferry the completed Pegasus vehicle to other global locations for aerial launch as needed.

The TriStar also carried the X-34 technology demonstrator built by Orbital Sciences starting in June 1999. The X-34 project was canceled by NASA in March 2001 before the vehicle could be launched from the TriStar mothership.

Chapter 8
SPACE SHUTTLE CARRIER AIRCRAFT

The moment of truth comes as the non-orbital Space Shuttle Enterprise separates from the descending 747 carrier aircraft high over the Mojave Desert on 12 August 1977. A strut extension put Enterprise at a slightly higher angle of attack relative to the Shuttle carrier aircraft (SCA) compared with Shuttles that were being ferried and not released. (Gene Furnish Collection via AFFTC/HO)

As the reusable Space Shuttle concept gained traction from 1969 onward, the logistics of moving orbiters expeditiously had to be considered. Although Vandenberg AFB on the California coast had been considered for some Shuttle launches and landings, only the NASA Kennedy Space Center in Florida would ever launch a Shuttle into space. Initially, all Shuttle landings were made on the 44-square-mile surface of Rogers Dry Lake at Edwards. This necessitated developing a way of transporting the 176,000-pound orbiters, spanning 78 feet and stretching 122 feet in length, from California across the United States to Florida for subsequent launches. Although the ambitious and optimistic rate of one Shuttle mission per week was never achieved, orbiters nonetheless needed retrieval to Florida after each flight.

C-5 Nixed for Shuttle Carrier

Wind tunnel tests validated the potential for mounting an orbiter on top of a modified 747 jumbo jet. (Wind tunnel tests were also made with a model of the huge expendable-liquid fuel tank that was part of the Shuttle launch stack, but these tanks were ultimately barged to the Kennedy Space Center.) NASA also considered using a Lockheed C-5 Galaxy as the means of piggybacking Space Shuttles, but preferred the low-wing configuration of the 747. In addition, for many years the Air Force's fleet of C-5s was considered a strategic asset, with guards posted wherever C-5s parked. The Air Force would have retained ownership of a C-5 Shuttle carrier, but NASA could own a 747 outright.

A NASA report on the Shuttle Approach and Landing Tests (ATL) elaborated on the thought that went into Space Shuttle mothership selection: "A multitude of ideas were evaluated. The extension of a large aircraft's landing gear, necessary to carry the Orbiter in an X-15 fashion seemed unreasonably complex. The idea of developing a new carrier, with the single purpose of carrying the Orbiter, was unreasonably costly. The options were reduced to carrying the Orbiter piggyback on either a Boeing 747 or a Lockheed C-5A.

"The technical concerns with both vehicles were related to clearances of the carrier vertical tail and relative aerodynamic effects during separation. The T-tail on the C-5A presented additional complications over the Boeing 747 aircraft. Of particular concern was the effect of the Orbiter wake on the C-5A T-tail immediately following separation.

"The actual decision to fly the Boeing 747 was based more on logistics than on technical rationale. The only C-5A available would have been loaned to NASA by the Air Force. Since the Air Force could recall the airplane at any time, NASA would not be able to schedule operations without risk."[104]

In addition to ferrying Shuttle orbiters from one site to another, the proposed Shuttle Carrier Aircraft (SCA) would be employed in the pre-orbital ALT in which an unpowered Shuttle (*Enterprise*, the only Shuttle that never entered space) would be released from the carrier aircraft high over Edwards Air Force Base to validate and explore the Shuttle's steep landing approach.

A 1974 comparison of 747 and C-5A characteristics found that the ability to clear the vertical tail of either mothership candidate would have a greater margin of safety if the orbiter were mounted at a higher angle of incidence than would be the most efficient for cross-country carriage of an orbiter. Moreover, if the SCA cut its thrust in a steep descent for separation, tail clearance would not be a concern. The C-5A's T-tail and high wing were both believed likelier to cause wake issues for the two machines during ALT separation flights, including the possibility of C-5A pitch-up if the Shuttle's wake impinged on the C-5's T-tail horizontal stabilizer and elevators.

Although both carrier aircraft candidates were believed to be capable of doing the job, some early evaluators favored the low-wing low-tail configuration of the 747, coupled with the overall availability of a 747 that could be owned by NASA and not shared with the Air Force.[105]

A former American Airlines 747 was bought by NASA on 19 June 1974, in anticipation of Shuttle duty. Initially it flew wake turbulence studies before being needed for the Shuttle.

Accordingly, in 1976–1977 Boeing modified this NASA 747 in Seattle with special dorsal mounting points to carry a Shuttle orbiter. This 747 was instantly recognizable by the two huge rectangular vertical fin endplates mounted to the

Concerns over the buffeting the aft end of a Shuttle could induce on the tail of the 747 led to the use of the streamlined shroud for early captive-carry and release flights during the ALT conducted above Edwards Air Force Base in 1977. (Gene Furnish Collection via AFFTC/HO)

The first mated flight of Shuttle and carrier aircraft took place on 18 February 1977, taking off from Edwards. The angle of attack for Shuttle mounts in the ensuing ALT of 1977 was steeper than later when Shuttles were simply transported from place to place on the 747 without an aerial separation. (NASA)

tips of the horizontal stabilizers to provide greater stability in flight with an Orbiter aboard. It carried the NASA registration N905NA.

When NASA engineers contemplated using the piggyback Shuttle-and-747 configuration to air-launch the Shuttle *Enterprise* for glide tests, they were pleased to find vintage French motion picture film of piggyback motherships launching research vehicles, as noted in a report: "Films of the flight of the French configuration were made available to NASA. Of interest was the relative incidence angle, the attach structure, and the pitchover maneuver to achieve separation. All were very similar to the design selected for the ALT program."[106]

While still bearing the red and blue paint bands of its former airline, the initial 747 SCA embarked on a unique phase of its service as a Shuttle carrier on 18 February 1977. The 747/Shuttle combination ultimately saw the Shuttle *Enterprise* launched from the 747 for glide tests over Edwards Air Force Base.

With mounting struts providing a different angle of attack than those intended for cross-country Shuttle ferrying, 747 SCA N905NA went aloft with the non-orbital *Enterprise* for the first captive-carry flight to see how the combination fared. Quietly fearless test pilot Fitzhugh L. "Fitz" Fulton Jr., whose skills with large test and mothership aircraft had been demonstrated time and again at Edwards Air Force Base, flew the initial 747/Orbiter *Enterprise* SCA lash-up that day. Veteran NASA pilot Tom McMurtry joined Fulton as well as Vic Horton and Skip Guidry on that 2-hour flight to check out handling qualities and low-speed flight characteristics.

For the first five 747/Orbiter flights the Orbiter was unmanned while the flight characteristics of the combination were explored and validated. A NASA report explained the process: "The ultimate aims of the flight test program were to certify the Orbiter/SCA configuration for ferry flight and to test the Orbiter approach and landing phase. In order to flight test the Orbiter, a separation maneuver was required and an initial part of the flight test program was designed to ensure that the separation was viable."[107]

The fifth captive carry flight yielded important information, as described in a NASA report: "The fifth flight of the inert series obtained data during two simulated launch maneuvers starting at ceiling altitude and terminating after

Chapter 8: Space Shuttle Carrier Aircraft

approximately 20 seconds of steady-state data following the 'launch ready' call by the SCA pilot. Both vehicles were configured as they would be for an actual separation with the exception of the Orbiter elevon. The elevon was positioned at –1 degree for emergency jettison for these early flights.

"An error in the SCA data base was discovered during these tests. The error was a result of the incorrect use of wind tunnel incremental data, and the aerodynamic data base was updated to the actual flight data. The inert tests verified that (1) the Orbiter/SCA configuration could achieve and stabilize on the separation parameters using the prescribed procedures without exceeding Orbiter or SCA constraints, (2) safe separation initial conditions could be achieved with the baseline separation configuration and airspeed, and (3) the mated configuration could recover from an aborted separation maneuver within the vehicle constraints."[108]

Flights six through eight saw *Enterprise* manned, but not separated from the 747: "Three captive-active flights were then flown with the Orbiter manned. The objectives of these flights were to verify (1) the separation configuration and procedures; (2) the integrated structure, aerodynamics, and flight control system; and (3) the Orbiter integrated system operations," the NASA report explained.

The envelope was expanding as the crew aboard *Enterprise* began making control inputs while still connected to the 747 SCA: "The first captive-active flight was restricted in airspeed and provided no separation data.

"The second flight included a full separation simulation. While the SCA maintained the separation conditions, the Orbiter crew moved the rotational hand controller (RHC) full forward and full aft to obtain elevon effectiveness data. Software limits restricted the elevon to move up 1.5 degrees and down 1.5 degrees from the 0-degree position for full RHC movement. Each position was held for 5 seconds to obtain steady-state data.

"Data from the load cells during this flight test were processed through the computer program to assess the elevon effectiveness. The results indicated a shift between the predicted values and the flight test data, equivalent to an approximately –1 degree bias in the Orbiter elevon position. Otherwise, the effectiveness of the elevon was in excellent agreement with preflight predictions.

"The third captive-active flight was a dress rehearsal for the actual separation. The elevon was moved from the climb position (–2 degrees) to the separation position (0 degrees) during the maneuver. The elevon bias did not appear during this test. This gave rise to questions regarding data repeatability and elevon position calibration accuracy. Fortunately, the first two separations were relatively insensitive to small elevon dispersions; *i.e.*, the 1-degree uncertainty still provided an adequate separation window. During the pre-separation maneuvers on these flights, more data could be obtained regarding the elevon bias for use in establishing separation conditions for more sensitive separations.

"To design the separation maneuver, off-line simulations were run to evaluate clearances and sensitivities. Manned simulations, for crew training, were made for the Orbiter and the SCA. In these manned simulations, the trainer vehicle, either Orbiter or SCA, was modeled to reflect the proximity aerodynamics. The SCA was modeled as the mated vehicle until separation and then was influenced by predefined proximity aerodynamic effects as it was flown away from the Orbiter. The Orbiter flew a predefined profile to the separation point. After separation, predefined proximity aerodynamics were applied while it was under the influence of the SCA."[109]

At this point, the designers were confident that separation would occur safely. To secure top management buy-in, simulations were run with computer graphics to show what separation was expected to look like, based on available data. Actual separation high over the Mojave Desert awaited the Shuttle testers.

On 12 August 1977, the 747 SCA carried *Enterprise* aloft over Edwards Air Force Base for the first actual release and subsequent glide to landing of a Shuttle vehicle. At an altitude just above 24,000 feet and a speed of 310 mph, Shuttle pilots Gordon Fullerton and Fred Haise became the first to pilot *Enterprise* in free flight when explosive bolts released the attached Shuttle as the 747 descended.

Things don't always go as planned, however; a NASA report recorded: "The (separation) maneuver differed from planned due to a larger-than-expected Orbiter pitch-up rate immediately following separation. This was probably due to the fact that an onboard computer failed at the instant of separation and distracted the crew."[110]

In anticipation of long ferry flights while attached to the 747 mothership, the Shuttles were fitted with a streamlined tail cone weighing almost 4 tons, and making the

The first tail cone–off ALT with Enterprise *and the NASA 747 carrier aircraft came with Free Flight Four on 12 October 1977. This gave* Enterprise *the same relative aerodynamic characteristics of an orbital Shuttle on its return into the atmosphere. (NASA)*

blunt rocket nozzles at the back of the Orbiter more aerodynamic. It was felt that this streamlining, in addition to making the deadheading Shuttles more efficient, would also diminish wake turbulence that could adversely fatigue the tail structure of the 747 carrier aircraft. This streamlined tail cone was in place on *Enterprise* for all of the captive-carry ALT flights as well as the first three free flights.

On the fourth free flight, the tail shroud was off *Enterprise* and the first portion of the mission was filled with analysis of the performance of the combination aircraft in this configuration. NASA described it: "The Orbiter without the tail cone attached presented two major problems with the separation phase of flight. First, the increased buffet level could possibly result in an SCA cockpit environment that would make it impossible for the SCA to attain the specified target conditions. Second, with the removal of the tail cone, the change in Orbiter pitching moment required +7 degrees of down elevon, which was well outside the elevon range tested in the preceding flights. The SCA tail loads and climb performance degradation created by the increased buffet and drag levels, respectively, were also unknowns.

"A fourth captive-active flight was originally planned to investigate the flight envelope of the tail cone-off configuration but was deleted. The objectives of the canceled captive-active flight were combined with free flight 4 and were evaluated in the first half of the flight. The optimum incidence for tail cone off was 5 degrees as opposed to the 6 degrees for tail cone attached. However, to reduce the number of variables, it was decided to leave the incidence angle at 6 degrees.

"The first portion of the flight was dedicated to a real-time assessment of the buffet-induced loads and verification of the separation configuration and target conditions. A real-time GO/NO-GO decision for separation was based on load cell data telemetered to the ground and displayed on strip-charts in the Dryden Flight Research Center control room.

"The buffet levels were determined to be acceptable from takeoff to maximum airspeed and a separation rehearsal maneuver was initiated. Had the data not matched the preflight predictions, a second rehearsal would have been flown to obtain elevon effectiveness over the untested range. The data in the first rehearsal, with the elevon deflected to +7 degrees, confirmed the preflight predictions and all parameters were within the acceptable separation window. A real-time decision was made to continue with the actual separation maneuver. Again, post-flight analysis in off-line simulations agreed well with actual flight data."[111]

In 1982, NASA explored the possibility of leasing the red-and-white prototype 747 used by Boeing. In 1983, proponents in NASA put forth arguments for acquiring a second 747 Shuttle carrier aircraft, considering using the aircraft as a carrier of B-1 bomber sections when not engaged in Shuttle ferrying. Consideration was given to making the 747s air-refuelable. The refueling concept could have alleviated a problem of anticipated gross weights higher than the typical Orbiter weights, allowing the combination aircraft to take off with fuel tanks only partially filled.

One refueling concept would have placed a receptacle on the nose of the 747, although the wake caused by a tanker so close to the 747-and-Shuttle combination was worrisome. A more radical refueling technique would have the 747 trail a hose, to which a tanker would plug in from behind. The cost of modifying the 747s for either style of refueling would have been in the millions of dollars, and the idea was shelved.

A second SCA was acquired in 1988. It was a former Japan Airlines 747-100SR, re-registered N911NA. With the end of the Space Shuttle program, both NASA 747 carrier

Chapter 8: Space Shuttle Carrier Aircraft

Space Shuttle Columbia rides atop a NASA 747 Shuttle carrier aircraft for the long haul from Palmdale, California, to Kennedy Space Center in Florida on 1 March 2001. When ferrying a tile-sheathed Space Shuttle, it was necessary to avoid rain showers because at airspeed, raindrops impacting the tiles could do damage. The flight from California to Florida required at least one fuel stop. (NASA Photo by Jim Ross)

The symmetry of shapes and the contrasts of forms is captured in this top-down view of a NASA 747 carrying Columbia back to Kennedy Space Center in 2001. Although the Space Shuttle was mounted for ferrying in a way that theoretically contributed some lift, this was hardly of benefit. The weight and drag of the orbiter atop the 747 required the normal fuel burn of 20,000 pounds per hour to double when toting a Space Shuttle cross country. (NASA)

On 28 September 1995 the Space Shuttle Discovery hitched a ride back to the Palmdale, California, plant where it was built, and where modifications were made. Palmdale kept a tubular truss crane available for such actions whenever a Shuttle needed to be mated or demated from its carrier aircraft. A more robust Mate-Demate Device (MDD) served Shuttle operations at the Dryden Flight Research Center on Edwards Air Force Base. (NASA)

Desert lightning punctuated the night sky as crews worked all day to prepare the Shuttle Discovery in the MDD at NASA Dryden Flight Research Center on Edwards Air Force Base on 14 August 2005. The Shuttle was slowly hoisted high enough to allow the 747 mothership to be towed under it for final mounting to the 747's special Shuttle carrier struts. (NASA Photo by Tom Tschida)

aircraft were retired in 2012. N911NA was slated to furnish parts as needed to support NASA's telescope-carrying SOFIA project 747; N905NA was put on display in Houston, Texas, with a Space Shuttle mock-up mounted to it.

Buran and Mriya for the Soviets

The United States was not alone in its pursuit of mothership capabilities for ferrying reusable spacecraft. The Soviet Union's *Buran* reusable Space Shuttle, looking startlingly like an American Shuttle orbiter, was carried on the back of the huge Antonov An-225 Mriya, a six-engine transport developed for that purpose.

Only one An-225 was completed; a second airframe was never finished. The An-225 fulfilled its promise of carrying the Soviet space shuttle handily. When that program was canceled in 1993 after only one unmanned flight of the Soviet orbiter, the An-225 served for a while as a military transport before lapsing into several years of storage. It emerged with Ukrainian civil registration to serve as the world's largest flying aircraft, dwarfing all but the museum piece Hughes Hercules, commonly known as *Spruce Goose*. The An-225, as this is written, regularly hires out to carry unusually large or heavy cargo.

The sole Antonov An-225 jet transport flew during the Abbotsford International Air Show in British Columbia on 13 August 1989. The An-225 was developed to carry the Soviet Union's intended Space Shuttle rival. With the demise of that program, the An-225, with part of its shuttle mounts in fairings atop the fuselage, has performed traditional heavy cargo lift missions. (Author)

This six-engine An-225 must have been the mother of all motherships, at least until the introduction of the twin-fuselage Stratolaunch carrier aircraft. The An-225 carried the Soviet space shuttle briefly and shows evidence of faired dorsal mounts in this 1989 photo taken at an air show in Canada. (Author)

Testbeds, Motherships & Parasites

Chapter 9
UNDERWING AND USEFUL

This ASM-N-2 Bat is mounted under the wing of a PB4Y-2 Privateer. (USN)

It started with the glide bombs tested and used in World War II, and it mushroomed in the post-war era. Twin- and four-engine mothership aircraft could carry everything from one-way glide vehicles to reusable reconnaissance drones beneath their wings. Glide bombs are on the discussion fringe of this book as they are more ordnance than parasite, but they are mentioned because they did hitch a ride and fly away from their bomber.

Glide Bombs Need a Bomber

Pioneer efforts in the United States included glide bombs released from variants of the pre-war Douglas B-23 bomber over Army Air Forces Muroc bombing ranges in the early 1940s. Refined iterations of glide bombs continued testing at Tonopah Army Airfield in Nevada later in the war.

The Bat could be carried and released by an SB2C Hell-diver, as seen in tests in 1946. (Navy/NARA)

The XQ-1 drone's vertical tail rested in a recess cut into the B-26 fuselage where a ventral power turret had been. This mothership was used for test drops at Holloman AFB when this photo was taken in September 1950. (NARA)

The XQ-1 target rode under the belly of a Douglas B-26 Invader when tested at Holloman, circa 1951. (NARA)

The XQ-1 drops from its B-26 Invader mothership for a launch over the vast New Mexico desert on 7 November 1950. (NARA)

The television-guided GB-4 glide bomb looked promising in 1944. Three B-17Gs (42-97518, 42-40042, and 42-40043) were modified to carry a pair of GB-4s externally under the wings and to house the necessary television and guidance equipment. Following testing at Tonopah, the special unit working on the GB-4s flew to Fersfield, England, where only aircraft 42-40043 made operational attempts at using GB-4s against German targets as part of Operation Batty.

Bat Goes to War

The U.S. Navy went operational with ASM-N-2 Bat glide bombs suspended from the wings of Consolidated Privateer patrol bombers in the waning days of World War II, identifying the Bat-capable Privateers as PB4Y-2B versions. Post-war tests included an Avenger and a Helldiver as launch aircraft for the Bat.

Post-war Navy testing used some PB4Y-1 Liberators as well as Privateers to launch underwing missiles for a variety of programs.

Radioplane Q-1

The black-bellied B-29 mothership mounted an experimental Radioplane XQ-1 from its right wing at Holloman on 3 August 1950. (NARA)

The post–World War II jet age created an immediate need for anti-aircraft

Chapter 9: Underwing and Useful   119

defenses commensurate with the new generation of fast movers. This also caused the Air Force to seek jet-propelled target drones to stand in as realistic bogeys. Not all of the drone projects of the late 1940s bore operational fruit. The Radioplane XQ-1, initially powered by a Giannini pulse jet, was intended for either ground launch with JATO and a mobile launch catapult or air launch. A contemporary Air Force description of this referred to a gender-neutral "parent aircraft" instead of a mothership.[112]

Perhaps because its ground-launched version would have no restrictions on vertical fin height, the original XQ-1 had a tall single tail that required special accommodations when carried by a mothership. A special Douglas B-26 Invader (43-22494) was modified with a ventral opening to house the tall tail inside the bomber's aft fuselage when the XQ-1 was mounted to its belly for test flights at Holloman AFB, New Mexico, in 1950–1951. Alternately, a geometric tube truss suspended from the left wing of a B-29 (probably 45-21748) provided clearance for the XQ-1 during test flights.

Parachute recovery was designed for the XQ-1. The Q-1 series evolved, replacing the pulse jet with a turbojet and shortening the vertical fin height by using two verticals instead of one. But it did not achieve production, as the Ryan Firebee became the service's choice for a jet target drone. The Radioplane Company was acquired by Northrop, which developed the Q-1 into the GAM-67 Crossbow anti-radar missile that could be carried by aircraft such as the B-50 and B-47. The Crossbow did not achieve production.

Firebee Finds a Home

The multitude of post-war underwing launch projects eventually distilled into the varied and successful Ryan Firebee jet drone. Versions were released from aircraft as small as a Douglas A-26 (and its Navy version, the JD-1). Later Firebees hit their stride while riding as many as four at a time

This early Q-2 Firebee test over Holloman, circa 1953, shows the drone dropping from the modified B-26 Invader bomb bay. Later, Firebees were underwing riders on aircraft ranging from Douglas B-26s to C-130s. (NARA)

under the wings of modified DC-130 Hercules transports that served as launching motherships. As early as 1953, a Firebee was carried and launched by an Air Force A-26 (B-26) Invader over New Mexico.

A 1956 Characteristics Summary for the Q-2A Firebee target drone described its relationship to its mothership: "Air launch from wings or bomb bay of B-26 launch aircraft." The method of carriage and release was to "Air launch from S-3 bomb shackles."[113]

For its early use as a target drone, radar signature masking was not an issue, and early Firebees showed up readily on radar. But increased risk on manned photo reconnaissance assets (including U-2s and jet fighter adaptations such as the RF-101 and RF-4), due to increased shoot-down capability by potential adversaries, prompted reconnaissance planners to consider reconnaissance drones evolved from the barrel-shaped Ryan Firebee. The reconnaissance drones are still widely called Firebees, but that name really originated for the predecessor target drones.

This Douglas B-26C Invader of the 3205th Drone Group carried a Firebee under its right wing in this 1955 photo. (AFHRA)

Several aspects required attention. Some radar cross-signature shrinkage was achieved with such items as radar-absorbent coatings and changes in the geometry of parts of the airframe. The Firebee was on its way to becoming a valued reconnaissance partner in the 1960s.

It was clear that a twin-engine carrier plane such as the Douglas A-26 would be inferior to a long-ranging and larger asset such as the Lockheed C-130. Lockheed DC-130s became the motherships of choice, able to carry as many as four Firebees, whereas the A-26 typically only hefted one of the drones. In service, it was typical for DC-130s to carry only two AQM-34 drones, with one the primary mission vehicle and the second a back-up contingency craft.

DC-130s could carry the drones to the launch point, but recovery methodology was initially problematic. Parachute recovery could lead to landings away from the designated landing area, and airframe damage was a possibility. Mid-air snatching of parachute-borne drones by helicopters became the answer.

The motherships of choice were 16 C-130 Hercules transports adapted to carry as many as four of the Ryan remotely piloted vehicles. Earlier, Douglas A-26s (and Navy JD-1 variants), as well as Navy DP-2E Neptunes, had carried Firebees aloft for transport and launching. The DP-2E could carry two Firebees.

The first two Hercules carriers were C-130A-models 57-0496 and 57-0497. They served as motherships for the Northrop Q-4 drone project (which also hitched a ride aboard a B-50 mothership). These two C-130s flew test missions with Ryan AQM-34 Lightning Bug and Fire Fly drone vehicles. They also launched the first operational Model 147B reconnaissance drone sorties over China.[114]

Some of the drone-carrying C-130s were labeled GC-130; others, DC-130. The G-prefix letter was instituted by the Air Force in 1948 to designate the carrier of a parasite airplane; hence, GRB-36 FICON carriers for RF-84K jet aircraft. The D-prefix letter indicates a drone director or drone controller. In this sense, a DC-130 is a drone director that also happens to carry its drones. In other instances, aircraft such as the redesignated DB-17 directed other drone B-17s but did not actually carry drones.

The Air Force accelerated its operational drone capability in the mid-1960s by modifying three more C-130As

Chapter 9: Underwing and Useful

A Firebee target named "Looey" grinned from the left wing of this Air Force B-26 drone carrier. Another Firebee is barely visible under the right wing as well. (Warren Bodie Collection via Peter M. Bowers)

(56-0514, 56-0527, and 57-0461) and three C-130Es (61-2368, 61-2369, and 61-2371) into DC-130 configuration. These were the 11th, 12th, and 14th E-models built by Lockheed-Georgia. The weight of operational reconnaissance drones was greater than that of earlier Firebee target drones, so the A-model C-130s carried only two recon drones. As DC-130Es became more abundant in the Air Force, some of the older A-models were transferred to the U.S. Navy for noncombat use.

The DC-130 accommodated stations and operators to start the AQM-34 engine in flight, ready it for launch, and once launched, the drone crew could track and control the remotely piloted vehicle.

The first operational mission launching an AQM-34 was on 20 August 1964. The AQM-34 overflew China. Over the next 11 years, DC-130s launched Lightning Bugs on 3435 sorties in support of the war in Southeast Asia.

AQM-34s could be configured to perform photo reconnaissance as well as communication and electronic intelligence gathering (COMINT and ELINT, respectively). The tasks of the unmanned Lightning Bug could include leaflet dropping and decoy operations. Chaff and electronic countermeasures could be carried by an AQM-34 to assist in the ongoing chore of foiling enemy air defenses.[115]

Air Force reconnaissance assets, including the manned U-2 and SR-71 aircraft, were under the umbrella of Strategic Air Command, as were the operational reconnaissance AQM-34s for the duration of the war in Southeast Asia.

An operational AQM-34 reconnaissance sortie saw the DC-130 takeoff in the morning from Kadena Air Base on Okinawa with a Lightning Bug under each wing. During midday, the DC-130 launched one AQM-34 on a 2- to 3-hour pre-programmed mission, while the DC-130 flew back to Kadena. After gathering the intel for which it was launched, the AQM-34 proceeded to Taiwan or some other friendly turf where it returned to earth under the canopy of a parachute.[116]

Initial missions occasionally saw Lightning Bug drones come down in less-than-ideal terrain, sometimes contested by the enemy. Extraction of the drone was thereby made more difficult, and damage to the reusable airframe could occur on landing or recovery. This prompted the development of specialized parachutes to lower the AQM-34s and traps beneath CH-3 helicopters to snatch the descending drones in flight. The recovery snare on the helicopter was the Mid Air Retrieval System (MARS).

The main drone parachute was surmounted by a smaller parachute riding high above it on a line. Two poles extended down and aft from the CH-3, with cabling between the poles designed to ensnare the upper parachute as the CH-3 pilot flew directly over the descending drone. The drone was then winched closer to the belly of the CH-3 for the ride back to a friendly airfield where the reconnaissance package was removed and flown by courier to Strategic Air Command headquarters at Offutt Air Force Base near Omaha, Nebraska. Meanwhile, another C-130 airlifted the drone to Kadena and back into the pool of available AQM-34s.

During a particularly busy time for drone missions in 1972, one CH-3 crew effected four MARS pickups of Lightning Bug drones in a single day.

In 1976 the operational DC-130s transferred to Tactical Air Command. After their combat operations were finished, some DC-130s were reverted to standard transport configuration and some continued as test motherships. In July 1976, a natural metal finish DC-130H took off from Edwards, mounting four heavy drones under its wings weighing a total of 44,510 pounds, more than 22 tons. That tonnage was believed to be a record for the amount of external weight hefted by a turboprop aircraft at the time.[117]

Here is USAF DC-130H (65-0979) with underwing drones. On the number-1 pylon (outboard left wing) is an AQM-34M, a photo-reconnaissance drone, equipped with two Sargent Fletcher 87-gallon external fuel tanks. Under the right wing of the DC-130 are two AQM-34V electronic warfare/jammer drones. The jammer drones likely preceded other aircraft into a target area. (AFTC/HO)

Chapter 9: Underwing and Useful 123

Chapter 10
SCALED COMPOSITES' MOTHERSHIPS

White Knight and Space Ship One are paced over the Mojave Desert by another radical design influenced by Scaled Composites, the Beech Starship pusher. (Courtesy of Scaled Composites, LLC)

The name Scaled Composites will forever conjure up images of aeronautical design genius Burt Rutan leading a sometimes unorthodox effort in the Mojave Desert to create one-of-a-kind airframes that reject convention while leveraging the immutable laws of physics. A proponent of everything from asymmetry to twin booms, crafted in smooth and curvy composite structures, Rutan and the company he nurtured have built or planned at least four unorthodox motherships.

Proteus Candidate for Space Launch

Proteus, a twinjet composite aircraft with unusual flying surface geometry, is capable of carrying a payload in the vicinity of 1 ton to an altitude of 50,000 feet, with flight duration up to 14 hours. It can swap shorter mission duration for heavier payload, and it can attain altitudes up to 60,000 feet. It derives power from two FJ44-2E turbofan engines that were modified by Williams International for

White Knight Two was on public display during the 2009 open house at Edwards. The development of this unique mothership characterizes an increasing movement by entrepreneurs to enter the world of space launch. (Photo by Tony Chong)

Scaled Composites' Proteus is said to have a space-launch mothership capability. This sunset photo was taken on 13 March 2002. (NASA Photo by Tom Tschida)

high-altitude performance. Moreover, although Scaled Composites says Proteus is a candidate as a mothership for space launch, it has proven its value as a testbed aircraft, often carrying specialized centerline pods for test work.[118]

White Knight Rewrites Mothership Rationale

Vital to Scaled Composites' successful winning of the $10 million Ansari X-Prize for privately funded spacecraft in 2004 was the original White Knight mothership. If most motherships have been frugal adaptations of existing airframes, White Knight was designed from the ground up for carrying the smaller *Space Ship One* aloft and releasing it for entry into near space. In addition to carrying *Space Ship One* ventrally, White Knight (the name is an homage to X-15 pilots Bob White and Pete Knight) uses an identical cockpit pod to that of the space ship. In this way, White Knight capitalized on design features including large spoilers to enable it to mimic atmospheric portions of the *Space Ship One* flight envelope. The cockpit of White Knight One is centrally mounted on the wing; two outboard booms host landing gear and twin T-tails. The gap between the tails gives additional freedom in the type of loads that White Knight One can carry and release.

The Model 318 White Knight by Scaled Composites is that company's innovative carrier aircraft for its Space Ship One *suborbital space vehicle. As a purpose-built mothership, White Knight's design from the outset accommodates the carrying and release of the smaller vehicle. Here, the mated pair is in flight near their homeport of Mojave, California. (Courtesy of Scaled Composites, LLC)*

126   Testbeds, Motherships & Parasites

In the year of the centennial of powered flight, Proteus gives visual confirmation of the aeronautical changes that have happened in 100 years. This photo was taken on 27 March 2003. (NASA Photo by Carla Thomas)

The fantastic purpose-built mothership White Knight Two attended the Edwards Air Force Base open house in 2009. Four PW308 turbofans power the aircraft, which spans just over 141 feet. (Photo by Craig Kaston)

White Knight can exceed 53,000 feet altitude, powered by two afterburning J85 jet engines. Its normal 82-foot wingspan can be modified to become 93 feet to enhance its climbing ability. White Knight can carry and release payloads weighing up to 7,000 pounds. In addition to the successful and repeated launches of *Space Ship One,* White Knight is positioned to be a mothership launch vehicle for what Scaled Composites terms "micro satellites."[119]

White Knight Two

Scaled Composites' stepping-stone approach to non-governmental space flight led to the development of *Space Ship Two,* a larger vehicle. This demanded an all-new mothership, White Knight Two, sometimes known as Eve. With a wingspan of 140 feet, White Knight Two is, as of this writing, the largest all-carbon-composite aircraft flying. (Other large aircraft known for their use of composites are not as fully invested in the carbon-composite structure as is White Knight Two.) The first flight of White Knight Two was 21 December 2008.

Four Pratt & Whitney PW308A jet engines propel White Knight Two. Unlike White Knight One, this larger mothership has two complete fuselages with crew compartments in each, culminating in tails with separated horizontal surfaces. White Knight Two is designed to be able to carry loads 30 percent greater than a full-up *Space Ship Two.*[120]

Stratolaunch Promises More

Under development by Scaled Composites for Paul G. Allen as this is written, the huge Stratolaunch aircraft system will be the largest aircraft in the world in terms of wingspan, at 385 feet. To simplify development, two 747-400 jumbo jets have been cannibalized for engines, landing gear, and portions of the cockpit. The Stratolaunch mothership will be powered by six 747 engines.

Stratolaunch and Orbital ATK announced in October 2016 their intent to use the huge Stratolaunch aircraft to carry as many as three Pegasus XL rockets for air launch of low-earth-orbit satellites weighing up to 1,000 pounds. With other mothership aircraft, versions of the Pegasus have already successfully placed more than 80 satellites in orbit.

Chapter 11
AIRFRAME MODS AND *SYSTEMS TESTS*

New large wings of the X-21A contained thousands of minute openings in upper and lower surfaces to suck air, keeping surface flow more attached and less turbulent. Underwing pods housed the suction generators. (Northrop via Tony Chong)

Of all the many aerial oddities shown in this book, perhaps none are more compelling as examples of the exceptional requirements of flight test than existing airframes modified with different wings, engines, and even cockpits to fulfill their experimental flight test roles. From a swept-wing World War II piston-powered fighter to a clipped-wing ground-bound landing gear test mule, or even a jet fighter fitted with a turbojet-powered supersonic propeller, airframe modifications produced some of the more unique aircraft ever flown.

Vega Virtues

In 1957, General Electric bought an old all-wood Lockheed Vega single-engine cabin aircraft in Texas and returned it to the GE flight facility at Schenectady, New York, for refurbishing and modification. GE's fascination with the vintage Lockheed was not rooted in a sense of history. The company needed a largely non-metallic testbed for research into radar countermeasures technology, and the 1929 Vega offered a solution.

This wooden Lockheed Vega made a valuable testbed for General Electric when an airframe without large amounts of metal was required for radar research. (General Electric)

Modified into a single-seater, this Fairchild PT-19 open-cockpit trainer was used by Boeing to help validate the design of the much larger B-29 Superfortress by flying with quarter-scale B-29 wings and tail surfaces. (Peter M. Bowers Collection)

The GE Vega testbed, filled with equipment, rambled through the skies over Schenectady as well as the Air Force's laboratories at Rome, New York, and Dayton, Ohio. For four years, until 1961, the vintage Vega served as a testbed. A year later, it was sold to antique aircraft restorer and collector David Jameson of Oshkosh, Wisconsin, who meticulously restored it and painted it to look like Wiley Post's famed Vega nicknamed "Winnie Mae." After Jameson's ownership, collector Kermit Weeks acquired the Vega and moved it to Florida.

One-Fourth as Big as a B-29

In a quirky twist on the theme of bombers as testbeds, Boeing employed a small single-engine Fairchild PT-19A trainer (41-20531) as an aerodynamic testbed for the B-29 Superfortress. The PT-19 was modified to have a single cockpit in addition to large quarter-scale versions of the B-29's graceful high-aspect-ratio wings and typical rounded Boeing-style fin and rudder.

Boeing initially had its hands full in convincing the Army Air Forces that the B-29 as conceived by the company was the right size for the job. It had higher wing loading than contemporary bombers, and some pressure was exerted to try to persuade Boeing to make the wing bigger to lower the wing-loading factor. Boeing successfully argued against this, and the B-29 wing underwent an atypically high amount of wind tunnel testing as well as flight testing on the unusual PT-19 testbed.

Clipped-Wing Invader

Real data was lacking on how aircraft tires interacted where the rubber met the runway. The answer was an instrumented

Douglas B-26B Invader (44-34137) with a huge modification; its wings were removed just outboard of the engine nacelles. The goal was to have a realistic 1-g weight on the wheels, uninfluenced by wing lift at speeds over 100 mph. Even the inboard wing remnants had spoilers installed to kill lift. With this aircraft, nicknamed "Wingless Wonder," it was possible to measure the sliding coefficient of friction between an aircraft tire and the runway. High-speed cameras captured the action.[121]

Strain gauges on the left main gear captured both drag forces and vertical forces, recorded on an oscillograph machine. The high-speed skidding tests revealed rubber skinning off the tire, liquefying, and vaporizing from the stresses of the skid on the runway.[122]

An anecdote about test flying at Wright Field indicates that another Wingless Wonder was made from an A-20, circa 1944–1945, and used on the test facility's special accelerated downhill runway for braking tests.[123]

Swept-Wing Kingcobra

American military services as well as the NACA took on a sense of urgency in the early post-war years as they sought to understand and quantify aspects of swept-wing design and jet-powered aircraft. A variety of researchers in the United States embraced the advantages of swept wings for fast, efficient aircraft. Not universally known was the low-speed performance of such wings. The U.S. Navy was vitally interested in this; the introduction of early-generation jet-propelled fighters aboard aircraft carriers was already fraught with peril.

The smell of burning tire rubber and the heat of working brakes attended tests by the Wingless Wonder, a B-26 clipped to put all weight on the wheels for high-speed wheel and brake tests up to 150 mph, circa 1950. The high-speed skidding tests revealed rubber skinning off of the tire, liquefying, and vaporizing from the stresses of the skid on the runway. (USAF/NARA)

The Wingless Wonder Douglas B-26 with clipped wings to keep it from flying put all weight on the wheels for wheel, brake, and drag-chute tests up to 150 mph, circa 1950, at Wright-Patterson AFB. (USAF Photo/NARA)

The Bell L-39 was a modified P-63 Kingcobra with wings swept back at 35 degrees. A fuselage plug aft of the cockpit placed the tail surfaces farther back and gave the horizontal stabilizer a different angle of incidence. Low-speed handling characteristics were explored, using a variety of leading-edge devices to promote stable airflow over the wing and ailerons. A ventral fin of varying sizes also altered the L-39's handling. (NARA)

Stripes painted chord-wise on the wings of the L-39 helped visualize airflow when tufts were used. The swept wings did not accommodate retractable landing gear for these low-speed tests, so the main gear always remained extended; the nosewheel could be retracted. (NACA/NARA)

The Bell L-39 testbed flew with no wing dihedral and 35 degrees of sweepback. The three-blade propeller was cannibalized from a P-39 Airacobra. (NACA/NARA)

Slow spool-up speeds for early turbojets prompted the creation of the hybrid Ryan FR-1 Fireball, a 1943 design that used a GE turbojet for speed, plus a Wright R1820 radial engine turning a three-blade propeller for the all-important acceleration demanded on takeoffs and wave-offs. However, as turbojets rapidly improved in the late war years and early post-war period, the need for a hybrid power system diminished.

The undeniable benefits of swept wings remained; the Navy needed to know how they would perform at the low end of the speed range during aircraft carrier takeoffs and landings.

To plumb that region, Bell Aircraft, the U.S. Navy, and the NACA evaluated a special P-63 Kingcobra that had its wings swept aft 35 degrees. The resulting testbed was designated L-39 by the Navy.

The Navy and Bell evaluated the L-39s at Bell's facility between April and late August 1946. By that time, some core truths about the viability of swept wings and wing slats for upcoming Navy jet fighters had been established.[124]

After the evaluations at Bell Aircraft, in August and December 1946, L-39s numbers-1 and -2 migrated to the NACA's Langley, Virginia, test facility for more research and quantification in the detailed NACA style.

A test report from the NACA described the swept-wing L-39: "The airplane tested had a wing with a straight center panel and outer wing panels, which were swept back 35 degrees at the quarter-chord line (38.7 degrees at the leading edge).... The main landing gear of the airplane could not be retracted, but the nose gear was retractable."[125]

This unusual mismatch in landing-gear capability was the product of the modifications made to a straight-wing P-63

Chapter 11: Airframe Mods and System Tests

Kingcobra by adding wedges near the wing root and sweeping the outer wing panels aft 35 degrees. The main landing gear was splayed toe-out by the geometry of the sweepback, so the wheels were realigned parallel to the line of travel. However, this meant that the standard inward P-63 main gear retraction system did not work, so the main landing gear legs were left in the extended position on the L-39, with the main wheel wells in the underside of the wing skinned over. The L-39 used the fuselage of a P-63A and the outer wings of a P-63E.

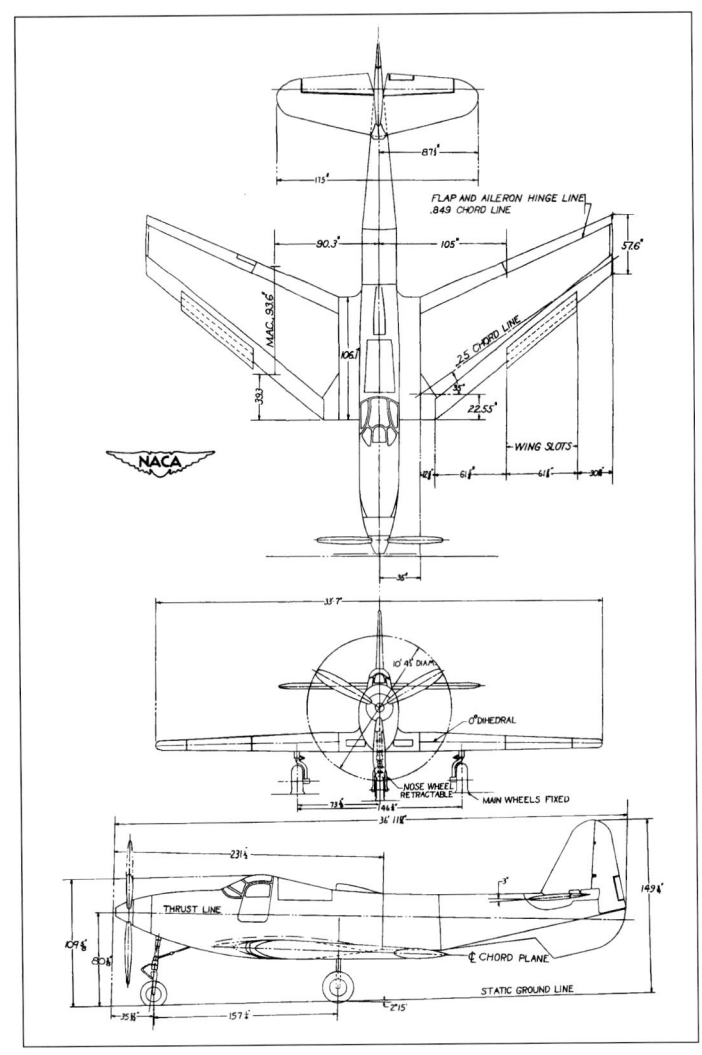

Three-view drawing of the swept-wing Bell L-39. (Courtesy of Craig Kaston)

Two L-39s were crafted, and they bore Navy Bureau of Aeronautics numbers 90060 and 90061. The design may have briefly been called P-63N by some, but officially it became the L-39. Unlike its P-63 origins, the L-39 used a three-blade propeller similar to that of the earlier P-39 also built by Bell. This use of a lighter-weight propeller may have been part of the effort to ensure that the swept-wing testbed had a proper center of gravity.

After initial flights in 1946, a 4-foot fuselage plug was fitted at the normal forward/aft fuselage production splice on the Kingcobra. The plug was shaped in a way that placed the L-39 tail farther aft and reduced the stabilizer's angle of incidence. The Kingcobra's use of tricycle landing gear, although unusual in a World War II fighter, was a bonus for a post-war testbed that was supposed to stand in for typically tricycle-gear jet fighters.

The swept wings of the L-39 were attached with zero dihedral; the upward sweep common to most aircraft of the era. In simplest terms, dihedral is a way to keep an aircraft stable in the roll axis; if it begins to roll toward one wing, the dihedral means that wing now has slightly more lift than the opposite wing because it is closer to horizontal, so the aircraft rights itself. Dihedral effect is the amount of roll moment produced by increments of sideslip.

For a swept wing at low flying speeds, dihedral effect on the L-39 became an issue. The L-39 exhibited negative dihedral effect at high speed and positive dihedral effect at low speed. The NACA report noted: "The [L-39] pilot considered the combination of high positive dihedral at low speed and negative dihedral at high speed particularly objectionable when occurring in the same airplane because he could not become accustomed to either condition."[126]

To induce rolling moment and sideslip that would produce dihedral effect for quantification, the L-39 sometimes was flown with one empty wing gas tank and the other one full, plus fuel in a nose tank. Some flights were made with a large ventral fin extension to compare performance of the aircraft with and without the ventral addition. The ventral fin extension increased directional stability of the testbed L-39. The decreased directional stability with the ventral fin removed made the L-39 more difficult to fly, in the opinion of its pilot, because this configuration made it easier to induce inadvertent sideslipping.[127]

In the L-39 configuration with 80-percent-span slots on the wing, the longitudinal stability was high throughout the speed range when the flaps were up. With flaps extended in this configuration, longitudinal stability was high at moderate speeds, but at a few MPH above stall, the stability decreased "and was neutral down to the stall," the NACA reported. The stalling characteristics of the airplane "were good with the flaps up or down when the 80-percent-span slots were on the wing. The airplane oscillated around all three axes at the stall. The attitude changes of the airplane were small during these oscillations and recovery from the stall could be made easily. With the ventral-fin extension removed, the amplitude of the oscillations at the stall increased rapidly."[128]

The L-39 was also evaluated with 40-percent leading-edge slots by the NACA at its storied Langley Memorial Aeronautical Laboratory on Langley Field, Virginia. Evaluators considered the possibility that the three-blade propeller mounted to the L-39 might cause different aerodynamic effects than would a jet engine. The NACA investigators reported their L-39 findings, saying, "In order to ensure that propeller operation would not mask any effects of sweepback, all the tests except aileron rolls were made with the engine idling. Aileron rolls were made with power for level flight as it expedited the tests, and power effects on the aileron rolling effectiveness were expected to be negligible."[129]

When the L-39 reports were published in 1948, the NACA had given a series of comparisons of the aircraft in different configurations of flap usage, ventral fin area, and wing slot amount and presence or absence. In some of the tests, the presence and change in span of wing slots did not contribute any notable differences in outcome, yet pilots were adamant that some configurations of the slots (or slats) were vital to tame the L-39's stalling characteristics.[130]

After L-39 testing was complete, Bell instituted some modifications to L-39-2 to make it more representative of the Bell X-2 experimental rocket aircraft for some company research.[131]

Up Periscope in an F-84G

Republic Aviation nurtured the potential Mach-3 XF-103 interceptor design for the Air Force in the early 1950s. The F-103 was envisioned as a speedster that would use missiles

With a one-eyed Cyclops artwork character, this unusual housing replaced the forward windscreen on a testbed Republic F-84G Thunderjet at Edwards in 1955. It allowed testing of a periscopic view for the pilot in anticipation of using a periscope on Republic's Mach-3 XF-103 interceptor, an all-titanium aircraft that was proposed but never built. (AFFTC/HO)

to defeat inbound Soviet bombers. The Air Force wanted the F-103 to have an unbroken cockpit upper surface to enhance streamlining, so the concept of a periscope for pilot vision was embraced; perhaps not universally but sufficiently to warrant a test of the concept.

A 1952 paper written by two researchers from Illinois University at Urbana teased the notion, saying: "Flight by periscope is practical and safe at least in slow-speed airplanes. It indicates that a radar or television method for presenting ground-reference information offers a practical aid to instrument flight."[132]

Republic tested a modified F-84G at Edwards to evaluate the periscope's functionality in a high-performance aircraft. The Air Force Flight Test Center historical records for July to December 1955 said that two flights with the unorthodox F-84 were made that September to "evaluate the Bausch & Lomb periscope installation. The pilot reported that the configuration as tested was not suitable for a high-speed tactical aircraft during inclement weather or night flying."[133]

It's a blustery, wet day at Lambert Field near St. Louis as the XF-88B spins its turboprop. The B-model was a well-crafted modification of the original pure-jet XF-88 prototype. (Dennis Jenkins Collection)

XF-88B Turboprop Testbed

The McDonnell XF-88 twinjet fighter of 1948 gained the name Voodoo, but no production contract. McDonnell revamped the design considerably, re-engined it, and emerged with the successful F-101 Voodoo several years later. With the two XF-88 prototypes out of work, the first one was modified by the addition of an Allison XT38 turboprop engine in the nose, complementing the original Westinghouse J-34 turbojets.

This version, the XF-88B (46-525), flew in 1953. It was retired in 1958. The NACA operated the XF-88B from its facility at Langley.[134] With its dual propulsion systems, the XF-88B could be seen in flight with its propeller shut down, the aircraft deriving power from the two aft-mounted J-34 turbojets.

The XF-88B program at the NACA facility was a joint Air Force-Navy-NACA venture. The Air Force furnished the XF-88; the Navy supplied the turboprop and turbojet engines; the research program was the responsibility of the NACA. The XF-88B was instrumented to quantify supersonic propeller performance. In some photos, measurement survey rakes may be seen on the nose of the fuselage, extending out both sides, behind the propeller. The goal was to determine the range produced by high-efficiency propellers for aircraft up to Mach 1.

The Allison XT38 turboprop engine in the nose is silent, its propeller blades feathered, as the XF-88B's two Westinghouse J-34 turbojets lift the testbed from the runway at St. Louis. (Dennis Jenkins Collection)

Potential military operational candidates for such turboprop systems were seen as long-range strategic bombers, long-range assault transports, tankers, and maximum-endurance tactical fighter-bombers. In addition, passenger transports might benefit from this technology.[135]

The XF-88B could attain level-flight speeds in excess of .9 Mach; it could exceed the speed of sound (Mach 1.0) in dives. The NACA described the propeller drive system of the testbed XF-88B: "The engine that drives the test propellers is an Allison XT38-A-5 turboprop engine, which has the commercial designation of Model 501-F1. This basic powerplant and later versions are used in several present-day airplanes. . . . The special gearbox provides two propeller rotational speeds of 1,700 and 3,600 rpm at a power section rotational speed of 14,300 rpm. Either of these propeller rotational speeds can be made available by selection of gear sets during the gearbox assembly."[136]

Aeroproducts made a supersonic propeller used on the XF-88B, and described by the NACA: "The supersonic propeller is a model of the Aeroproducts 12-foot-diameter propeller that was designed to absorb 7,500 hp in its full-scale version. The propeller is a three-blade configuration with a 7.2-foot diameter. The blades were fabricated from solid steel having an ultimate tensile strength of 180,000 pounds per square inch."[137]

The NACA crew at Langley did what they could to maximize available fuel in the experimental XF-88B for flight tests: "The normal flight procedure is generally as follows: The airplane is towed to the end of the runway, clearance is obtained from the tower, and the J-34 jet engines are started, checked, and afterburners fired for takeoff. In order to conserve fuel the afterburners are cut off and the T-38 turboprop is started at an altitude of 5,000 feet to assist in the climb.

"Level-flight Mach numbers up to approximately .9 Mach can be obtained by using both the J-34 turbojet engines with afterburners and the T-38 turboprop engine. Mach numbers up to and slightly above 1.0 can be obtained in pushovers to dives of 30 degrees from 40,000 feet.

"Data are continuously recorded as the pilot accelerates from .5 Mach number up to the maximum test Mach number."[138]

Aeronautic researchers are at risk when they propose a new research program with lead time. By the time the program can be executed and the data reduced and analyzed, the original reason for conducting the research may have been made moot by newer findings and alternate capabilities. The NACA lobbied for a high-speed propeller research aircraft. The XF-88, converted and ready for use by 1953, fit the bill. But the aviation industry grew cooler toward the notion of high-speed turboprop aircraft even as NACA research was in planning and under way.

The notorious issues attending the XF-84H at Edwards in 1955 and 1956 probably did not help the cause of high-speed propeller advocates. Even the NACA began dismantling its propeller research subcommittee. So although the work performed by the XF-88B team was groundbreaking, the findings went largely unused.

In May 1959, the new NASA, inheritor of the NACA's roles and responsibilities, published a memorandum on some of the final research done with the XF-88B. It was a project intended to quantify performance of a propjet designed for the slightly slower transonic speed around .82 Mach, and intentionally made for minimum noise. Although this propeller did achieve lower noise levels, it did so at a drop in efficiency. The near–Mach 1 propellers had yielded efficiency in the range of 97 percent; the .82 Mach airspeed of the new propeller was achieved with an efficiency of only 68 percent. The quieter propeller, in static and takeoff conditions, produced 117.5 decibels, which was about 5 decibels below the levels of the earlier propellers.[139]

Propfans made a brief experimental resurgence in the late 1980s as McDonnell Douglas explored newer designs that could be fitted to DC-9-style airframes, but the transport industry did not embrace the idea then either.

XF-84H Broke a Different Kind of Sound Barrier

The Cold War 1950s made Edwards the crossroads for all kinds of aircraft projects combining new engines and new aeronautical breakthroughs in the pursuit of both pure and applied research. The U.S. Navy and Air Force both expressed interest in a swept-wing turboprop fighter concept that became the Republic XF-84H. An Allison XT40A-1 turboprop engine mounted behind the pilot powered it; an extension driveshaft spun a three-blade supersonic-speed propeller fitted in the streamlined nose. The T40 also contributed thrust through exhaust from the tail of the XF-84H, and it was intended to incorporate an afterburner.

Poised with promise on Republic's ramp, the first of two XF-84H prop jets (51-7059) was sent to Edwards for flight testing. Although sometimes vilified for its noise and other problems, it is part of the role of a testbed aircraft to provide experimental data, positive or negative, which can inform future engineering and design choices. (Republic)

The second of two Republic XF-84H prop jets flew at Edwards, sometimes with a small Confederate flag in the rear cockpit window denoting test pilot Hank Beaird's Alabama roots. (AFFTC/HO)

The ambitiously powerful combination of the T40 and a powerful propeller led Republic to incorporate a triangular shark fin–like vane behind the cockpit canopy in an effort to tame torque from the propeller.[140]

The NACA ran tests on a Republic wind tunnel model of the then-unflown XF-84H at the Langley Memorial Aeronautical Laboratory's wind tunnel facilities. A salient change came from these model tests: The horizontal tail was relocated from a low cruciform location typical of other F-84s to a high T-tail perch that kept the horizontal out of the turbine's prop wash.[141]

Before the first flight in 1955, the Air Force Flight Test Center defined the upcoming XF-84H test program's goals: "To solve the problems involved in the use of high-powered turboprop engines in single-engine aircraft. Flight test data is necessary to validate previous static and wind tunnel test data. These XF-84H aircraft (S/Ns 51-17059 and 51-17060) are research vehicles for testing supersonic propellers. There are several propellers under development, which will be tested, and it is probable that these aircraft will be under test at the Air Force Flight Test Center for several years."[142]

Brig. Gen. J. Stanley Holtoner, then the Air Force Flight Test Center (AFFTC) commander, described the XF-84H program in August 1955: "XF-84H test objectives are: to obtain basic data required to design and develop supersonic propellers; to obtain solutions to the problems involved in the use of high-powered turboprop engines in fighter aircraft; and to conduct a 50-hour qualification test-stand run of the engine-afterburner-propeller combination."[143]

The XF-84H first taxied on 28 March 1955; loss of engine oil pressure curtailed

The XF-84H turboprop, seen here at Republic's Farmingdale, New York, plant, used a three-blade propeller with broad blades typical of the designs intended to convert turbine energy into aircraft thrust. (Republic)

gram. On that flight, the propjet attained a maximum speed of 260 knots indicated and an altitude of 20,000 feet. In August, as the XF-84H was being towed to the hangar, its nose landing gear collapsed and the propeller struck the ground. Tests were halted while modifications were made.[146]

A 1 August 1955, Project Assignment Directive set out a plan to measure jet engine noise at Edwards. Aircraft, including the turboprop XF-84H, were identified to participate in the noise surveys. From the AFFTC, test pilot Lt. Col. Frank K. Everest, Jr., was assigned to the project.[147]

The XF-84H was designed as an aircraft with high-subsonic flight speeds, but its propeller blades exceeded the speed of sound in rotation. The XF-84H gained a reputation for noisy operation that made this aircraft stand out and earning it the sometime-nickname "Thunderscreech." It has been reported that the XF-84s were eventually banished to the barren reaches of Rogers Dry Lake for engine runs, so noisy were their supersonic propellers.

Performance and handling issues were reported with the H-models, and by October 1956 their program was ended. XF-84H 51-17059, which made eight of the total 12 flights, was retired to an elevated display at the Kern County Airport in Bakersfield, California. In 1999, the national Museum of the U.S. Air Force moved this XF-84H to Ohio for restoration and display in the museum. The second XF-84H, 51-17060, made four flights before program termination. Its

this effort for two months while Republic worked on a fix. General Holtoner summarized the resumed taxi tests, which led to liftoff at an indicated airspeed of 160 knots: "A left-wing roll-off was experienced upon liftoff."

On the ground, rudder control came in for some criticism in a letter report from Holtoner: "In general, directional control using rudder is considered to be marginal when the throttle is retarded to the flight idle position at 100 knots and would result in loss of directional control using rudder control only; however, use of brakes provided adequate control."[144]

The T40 was tested in a ground stand with the afterburner in May 1955. An Aeroproducts propeller intended for flight testing was installed in June; some rework was found necessary during simulated flight tests.[145]

The first official flight of the XF-84H was on 21 July 1955, with a Republic pilot at the controls for the beginning of a company-run Phase I flight test pro-

The long bullet-nose spinner on the XF-84H is neatly faired to the circular cross-section of the fuselage. (Republic)

Chapter 11: Airframe Mods and System Tests

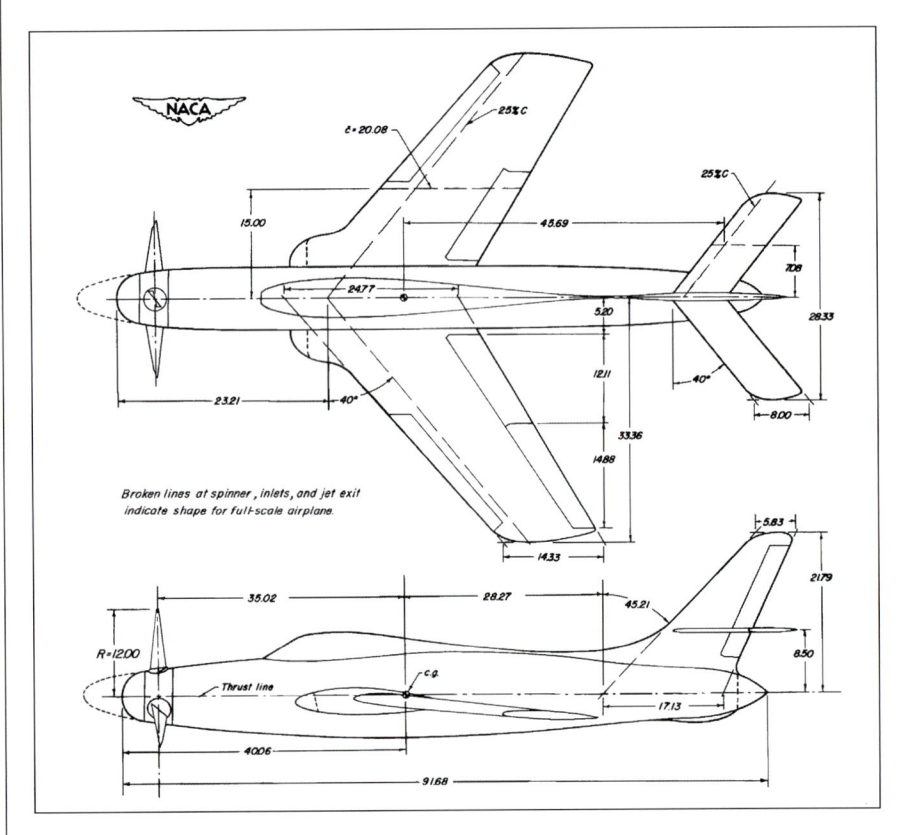

Three-view drawing of the Republic XF-84H showing contours of the nose section and extended "bullet" prop spinner. (Courtesy of Craig Kaston)

disposition is presumed to have been the salvage yard, with its T40 engine serving another aircraft development program.

Only two pilots flew the XF-84H; both were Republic test pilots. Henry "Hank" Beaird was the primary pilot and Lin Hendrix flew it once. Republic Chief Test Pilot Carl A. Bellinger was also at Edwards in the XF-84H era. In 1975 he recalled, "The noise level of the XF-84H, put out by the supersonic prop, was indeed murderous. In fact, we were barred from operating on the flight line due to the fact that the tower could not hear any transmissions while the H was running. We were assigned a spot about 4 miles away to do our run-ups. This noise level did not bother the pilots since it was confined to a 30-degree included angle in the plane of the propeller."[148]

XB-47D Tries Turboprops Inboard

The bloom of interest in high-speed turboprops in the 1950s even embraced bombardment aviation. Two sleek B-47B Stratojet bombers were modified to replace the paired, podded, J47 engines on their inboard pylons with a single Wright YT49 turboprop engine on those pylons, making a four-engine B-47 with mixed turbojet and turboprop propulsion.

The XB-47D concept design began in February 1951. By April, Boeing had a contract to convert two B-47Bs to D-model configuration. Research goals included a desire to see how turboprops fared on swept-wing aircraft, and to test the Wright T49 with a propeller combination. The T49 was an experimental propjet adaptation of the

The Boeing XB-47D turboprop was photographed at Boeing Field. The broad propeller blades are shown to advantage. Wright YT49 propjet engines replaced the paired J47 turbojets on the inboard pylons, making this the only four-engine B-47. (Boeing Photo via Tom Cole)

The XB-47D flew with two Curtiss-Wright T49 propjets. This engine model was tested on the nose of the company's 299Z five-engine B-17 testbed aircraft. (Peter M. Bowers Collection)

Wright J65 turbojet, which was an Americanized variant of the British Sapphire turbojet.

The J65 ultimately enjoyed success on aircraft such as the A-4 Skyhawk, B-57 Canberra, and F-84F Thunderstreak. If the purported long-range capabilities of high-speed turboprops turned out to make the XB-47D a practical bomber, so much the better. But even the pure research aspects of the turboprop Stratojet project held interest.[149]

The T49 was test-run in 1952; the first flight with the turboprop engines on the XB-47D was initially anticipated for some time in early 1953, but problems with the engine-propeller combination delayed this, as did shortages of government-furnished equipment for the project. The first XB-47D to fly was 51-2103 on 26 August 1955. The second XB-47D did not fly until 15 February 1956. The two propjet B-47s yielded research data, but no follow-on production ensued.[150]

The XB-47D achieved a top speed of 597 mph at an altitude of 13,500 feet.

X-21A Testbed Based on B-66

John K. Northrop's lifetime of design contributions focused on aerodynamic efficiency. His influences are evident in the smooth and strutless structure of the Lockheed Vega that he helped create for that company and his own lifelong dream of perfecting the flying wing attested to his interests.

Not surprisingly, even in the post–John K. Northrop era, the company bearing his name was engaged in probing sometimes radical means of achieving aerodynamic efficiencies. In 1955, Northrop used a bailed Lockheed F-94A jet interceptor for laminar flow studies. The goal was to baseline laminar flow numbers and to determine skin-friction drag. The evaluation and quantification with the EF-94A in 1955 was intended to inform a bigger laminar flow program to determine the optimum type of boundary layer control.[151]

That bigger program was the X-21A of 1963–1964. Sometimes a testbed can succeed in the goals its creators set for it, advancing aeronautical knowledge significantly, and yet its leading-edge technology may wither for a lack of practical application.

In 1960, Northrop began crafting two testbeds to validate concepts of laminar flow control (LFC) across a wing surface that were said to offer great drag-reduction efficiencies in cruising flight. Unlike straight wings, swept wings can cause more span-wise airflow with an associated turbulence that robs efficiency. Northrop set out to tame disruptive airflow by using thousands of tiny slots in the wing's surface, through which part of the passing airstream was sucked in by vacuum, thereby generating greater laminar adhesion and efficiency over more of the chord of the wing than if no LFC is employed.

In an effort to validate the concept, new swept wings with 816,000 metering holes and 68,000 ducts were fabricated at Northrop-Hawthorne in southern California. The wing surfaces had more than 3 miles of little slots sawed into the outer skin surfaces on top and bottom, from root to tip. The wings were mated to a pair of modified Douglas WB-66D fuselages (55-0408 and 55-0410) to create two new X-planes designated X-21A, and retaining the B-66 assigned Air Force serial numbers.

The changes were dramatic. The new wings spanned 93 feet 6 inches compared with only 72 feet 6 inches for standard B-66 wings.[152] Whereas traditional B-66s carried a J71 jet engine under each wing, the new test laminar wings were clean, except for a housing containing suction generators under each wing and extending to the rear. The B-66 fuselages were strengthened to take the weight of two aft-mounted, podded J79 jet engines.

The X-21A was on the cusp of the computer revolution. Boundary layer information was calculated on IBM computers using Fortran programming language. When the X-21 was tested in the early 1960s, transport aircraft were not routinely air-refuelable. Northrop engineers considered LFC to be a potentially beneficial range extender in the transport realm.[153]

A laminar flow control testbed was the X-21A, heavily modified by Northrop starting with a Douglas B-66. The highlight was the new wing, unimpeded by engine mounts. The power for the X-21A was relocated in two aft fuselage pods. (Northrop via Tony Chong)

In reality, this system of micro slots and suction, although interesting on a testbed flying in the dry air over the Mojave Desert, could run into operational degradation quickly. Snow, ice, rain, dust, and insect impingement all were potential degrading agents. This fact tended to place LFC in the realm of aeronautically curious phenomenon instead of being a practical solution to laminar flow improvement. Another development that lessened the need for LFC was the advancement of large turbofan jet engines that would efficiently carry transports farther.[154]

The first flight of an X-21A was 18 April 1963; the flight research program ended in 1964. In that time, the X-21As routinely demonstrated the ability to achieve laminar flow on 73 percent of the upper wing surface and 75 percent of the lower surface. This laminar flow increase over a non-slotted passive wing meant that an X-21A with 2 hours and 25 minutes of flying time with the system turned off could achieve 4 hours of flying with the system turned on, using the same amount of fuel in each case.[155]

The official Air Force Flight Test Center historical record for July to December 1963 gave its own account of the X-21A Category I flight test program: "The Norair Division of the Northrop Corporation modified two WB-66D airplanes to a Laminar Flow Control configuration, which were then designated X-21s by the Air Force.

"Norair modified the airplanes under Contract AF 33(600)-42052 for about $36 million including a flight test program supported by the AFFTC to determine the feasibility of incorporating Laminar Flow Control on future aircraft.

"Originally, AFFTC participation with Norair was scheduled to commence in March 1963 and then the AFFTC would pick up the full program with two aircraft in January 1964 for a test program anticipated to last about two years. The program slipped, however, and the first airplane (USAF S/N 55-408) never arrived at the Flight Test Center from the Norair Hawthorne, California, facility until 18 April 1963, and the second airplane (USAF S/N 55-410) was not ferried to the Center until 15 August 1963. Both aircraft underwent contractor airworthiness testing and laminar flow control systems flights throughout the remainder of 1963.

140   Testbeds, Motherships & Parasites

"Laminar flow control was a system designed by Norair to eliminate the major portion of air-friction drag that slowed an aircraft in flight. Norair's approach to the problem was to place row upon row of paper-thin slots span-wise along the upper and lower surfaces of each wing. The offending turbulent air was then drawn inboard through these slots by two Garrett AiResearch suction turbopumps and expelled back of the wings. Initial laminar flow control (LFC) flights encountered LFC pump and instrumentation malfunctions and metal shavings in the LFC wing slots.

"When these problems were overcome and LFC testing began, the test results were less than favorable in that laminar flow could be achieved only on the wing leading edges. Attempts to obtain full-wing laminar flow by progressively moving the laminar sensing probes toward the wing trailing edge only worsened the test results the farther back the probes were placed.

"Number 1 (55-408) airplane during a structural demonstration rolling pullout maneuver during late August incurred a structural failure in the aileron actuator connection member and in the primary engine support member. Nevertheless, Norair completed its contractor demonstrations of structural integrity, stability, and control on the two airplanes before the end of 1963. However, the laminar flow test results were so erratic and inconclusive that the contract with Norair was extended beyond the original 31 December completion date.

"Center test pilot and Chief of the Transport Branch under the Flight Test Operations Division, Major Joe S. Schiele, who flew the X-21 on 18 June 1963, was named AFFTC X-21 project pilot (additional duty) in mid-1963. Center project engineer monitoring the contractor Category I was Mr. Robert Sudderth, Performance Engineering Branch, Flight Test Engineering Division."[156]

Work on the X-21A program continued into 1964, during which time more promising LFC effects were achieved.

NC-131 Mimics Many

The world of testbeds is anything but haphazard. Sometimes, an airframe of convenience is modified to test a new component or system, but legitimate, safe flight test procedures demand thorough planning in the testbed process. Cornell Aeronautical Laboratory (sometimes known by the initials CAL and usually by the name Calspan) developed a thoughtful testbed aircraft that could mimic flight capabilities and situations for a wide variety of airplanes.

The genesis of this clever copycat was the unbuilt X-20 Dyna-Soar spaceplane. When the X-20 was under development in 1959, Calspan's Walt Breuhaus and colleagues in the industry posited the value of making an airborne inflight simulator to gain knowledge and training on the anticipated performance of the X-20. With the cancellation of the X-20 program in December 1963, the inflight simulator idea remained a spark. The proposed Boeing SST was a new contender for the inflight simulator, but it remained for the B-1 Lancer to bring the idea to fruition.

The Total In-Flight Simulator (TIFS) was a unique testbed with a complete second cockpit and the ability to mimic the performance of a variety of aircraft. (USAF)

Chapter 11: Airframe Mods and System Tests   141

The TIFS NC-131 had an alter ego with a different nose for radar and instrumentation testing. (USAF)

Known as the Total In-Flight Simulator (TIFS), its base was a much-modified turboprop Convair C-131. Its Air Force serial number was 53-7793; later it flew with civil registration number N793VS. Design work began in 1966 under the auspices of the U.S. Air Force. From 1967 to 1971, the simulation system was created and installed. The first aircraft simulated in flight by TIFS was the B-1 on 21 June 1971.

Key to the success and utility of TIFS was the fabrication of a complete second cockpit in the nose ahead of and below the normal C-131 flight deck. It's a philosophical line between testbed and simulator; a testbed traditionally evaluates hardware in the air; a simulator simulates flight conditions. TIFS simulated other aircraft, but in actual flight instead of a ground-bound simulator, sometimes blurring the demarcation between testbed and simulator.

A 1972 report by Cornell Aeronautical Laboratory (CAL) explained the promise of TIFS: "The Total In-Flight Simulator (TIFS) is a variable stability airplane, which has been developed by modifying a C-131 twin-engine transport. It has been designed to reproduce in actual flight the flying qualities of a wide range of large airplanes. There are several basic features that provide this capability. An evaluation cockpit has been added, which is entirely separate from the normal airplane command cockpit. With the addition of direct lift flaps and aerodynamic side-force surfaces, there is control not only of the moments about all three axes but also the forces acting along the three axes. An elaborate and versatile automatic flight control system has been installed to generate the required motions of the aircraft in all six degrees of freedom."

TIFS was instantly recognizable for its secondary cockpit and for tall vertical vanes on each wing, extending above and below the airfoil. The CAL report continues its description: "The TIFS airplane . . . is a modified C-131H, which is the military counterpart of the Convair 580. This version of the C-131 is equipped with Allison 501-D13 turboprop powerplants of 4,000 hp each. The zero fuel weight is in the range [of] 47,000 to 49,000 pounds with a takeoff maximum gross weight of 54, 600.

"The simulation cockpit is removable, so that it can be replaced by other cockpits of different configurations. The direct lift flaps extend from the inboard end of the ailerons to the engine nacelle and in that region replace the normal landing flap. Inboard of the direct lift flaps, the normal Fowler flaps for landing have been retained. The direct lift flaps are plain flaps with a total area of 108 square feet.

"The surfaces for generating aerodynamic side forces extend above and below the wing and are pivoted about an axis normal to the wing plane. Their total area is 100 square feet (50 square feet on each side of the aircraft), and they can be deflected plus-or-minus 30 degrees. The installations in the aircraft include sensors, computers, control actuators, displays, and flight data recording equipment."

The cockpit grafted onto the TIFS NC-131 was generic, with more expansive glazing than most aircraft. This was deliberate, the designers said, to allow portions of the windows to be masked off to represent the view from several specific aircraft cockpits.

The report encapsulated the heart of TIFS: "If the TIFS cockpit is to be made to move as required to duplicate the motions of another airplane, it is necessary to control all six degrees of freedom; that is, to have control over moments about, and forces along, all three axes. Through proper control of all six degrees of freedom, it is possible to simulate, for example, the motions of a large airplane's cockpit, which is located a great distance ahead of the airplane's center of gravity. The low lift curve slope of a delta wing airplane can be reproduced. The troublesome task of controlling sideslip in turn entries and recoveries for very large airplanes can be studied through use of the variable stability system."

The choice of the C-131 as host for the Total In-Flight Simulator aircraft was deliberate, as explained in the report: "In simulating another airplane, the basic vehicle characteristics should not show through and intrude upon the simulation. For example, if an airplane with flexible, swept wings had been chosen, the structural response to gusts would be that of the base airplane. It would be most difficult to suppress these natural response characteristics, let alone to try to simulate the structural response of another airplane. In general, the best simulation of other airplanes can be produced if the base airplane has fundamentally simple and straightforward characteristics. Specifically, it is highly desirable that it be relatively rigid and that it have uncomplicated stability and control characteristics."

NASA's F-8 Crusader testbed for the supercritical wing design was an aesthetic blend of new aerodynamics and classic Navy jet fighter, as seen in this 1973 photo. Rockwell International built the supercritical wing for the F-8 testbed under a $1.8 million contract. Whereas the original F-8 wing had a span of just over 35 feet, the supercritical wing spread to 43 feet. This testbed proved the efficiency of this modified wing, variations of which have served to improve the fuel mileage of modern jet transports. (NASA)

Initially, the developers of TIFS envisioned it as a way to simulate larger aircraft in flight. They said that it is easier to make a smaller aircraft such as the C-131 behave like a C-5 than it is to make a large aircraft behave like a smaller one. And, it is less expensive to fly and maintain a smaller machine.

Between its first simulation in 1971 and retirement to the National Museum of the United States Air Force in 2008, TIFS was an airborne stand-in for military, commercial, and NASA aerospace vehicles as diverse as the B-1, Space Shuttle, Concorde, Compass Cope (a remotely piloted vehicle), Tacit Blue, X-29, B-2, YF-23, C-17, C-141, IPTN N-250, X-40A, and Airbus A300 (to assist with a mishap investigation). TIFS gave several intended X-29 pilots the chance to appreciate the aircraft's flying qualities before they actually piloted the radical, forward swept-wing X-29. Vehicles such as the Compass Cope and X-40A were unpiloted, but the suite of computer flight controls aboard TIFS allowed the characteristics of these machines to be explored, with the safety redundancy of pilots in the TIFS cockpit.

TIFS also simulated the flying qualities of a number of aircraft that never left the drawing boards, including the unbuilt MD-12X wide-body trijet, Northrop Grumman supersonic tailless vehicle, and Boeing 7J7. Some TIFS flight programs were for generic notional aircraft listed simply as Twin Fuselage, High Speed Civil Transport, or Large Aircraft Flying Qualities.

NASA F-8 Supercritical Wing

When NASA wanted a high-performance aircraft to test a promising new wing design, the Vought F-8 Crusader seemed to be made to order. In its operational configuration as a carrier-based supersonic jet fighter, the Crusader had a high-mounted wing that changed the angle of incidence by rising above the fuselage at an angle. Wing replacement on such an aircraft was more practical than with traditional mid- or low-wing designs.

NASA's Richard Whitcomb had theorized a "supercritical" wing airfoil that originally was envisioned as a way to achieve efficiencies at near-sonic speeds around .98 Mach. Aerodynamic shaping bulges were added to the fuselage to further accommodate the high-speed efficiencies of this supercritical wing design. Whitcomb was right: Test data showed that the modified F-8 testbed increased its transonic efficiency by as much as 15 percent.

Rockwell International built the supercritical wing for the F-8 testbed under a $1.8 million contract. Whereas the original F-8 wing had a span of just over 35 feet, the supercritical wing spread to 43 feet.

The impact of fuel price increases around 1973 coincided with a decreased interest in commercial aircraft that could fly at near-sonic speeds, and increased the desire for greater fuel efficiency. An evolved supercritical wing shape helped achieve these goals and has been applied to a number of large air transport designs in the ensuing years.

The first flight of the NASA supercritical wing F-8 was 9 March 1971; the final flight was 23 May 1973. The modified F-8 Crusader was placed on display at the NASA Dryden Flight Research Center (later Armstrong Flight Research Center) at Edwards.

NASA F-8 Fly-by-Wire Testbed

A NASA research project that did much to ensure the benefits of digital fly-by-wire control systems involved a testbed Vought F-8C Crusader jet fighter. Instead of the Crusader's standard mechanical control inputs, the Digital Fly-By-Wire (DFBW) testbed relied on a spare Apollo digital flight control computer left over from that space program.

In 1972, NASA claimed this to be the first DFBW aircraft without a mechanical control backup, a significant step in building confidence in computer-based flight control systems. NASA's Gary Krier made the first flight on 25 May 1972. Flown at Edwards, home of NASA's Dryden Flight

The Digital Fly-By-Wire (DFBW) testbed F-8C Crusader was photographed in 1973. In 1972, NASA claimed this to be the first DFBW aircraft without a mechanical control backup, a significant step in building confidence in computer-based flight control systems. (NASA)

Research Center (renamed Armstrong Flight Research Center), the F-8 DFBW aircraft pioneered the way for computer control systems onboard aircraft such as the radical X-29, which was uncontrollable without the rapid input of its computers. This relaxed stability could improve maneuverability and efficiency, and has been a hallmark of operational DFBW systems ever since.

F-16 VISTA Inflight Simulator

The F-16 VISTA aircraft had input from Calspan to make it a research vehicle with variable stability and the ability to mimic the performance traits of other aircraft configurations. In the sometimes-tortured world of acronyms, by ignoring entirely the word "stability" in the name of this special F-16, someone in the U.S. Air Force derived the acronym VISTA from Variable stability In-flight Simulator Test Aircraft.

A Block 30 F-16D airframe (86-0048) with Block 40 avionics with a digital flight control computer was chosen to host the VISTA equipment. The airframe used for the F-16 VISTA has a long, boxy dorsal compartment similar to that found on some Israeli F-16s. This provides the space to house VISTA's unique equipment suite, enabling it to simulate the performance of other aircraft types. To broaden the jet's simulation capabilities, heavy-duty landing gear was installed to permit sink rates faster than normal F-16 landing rates.

VISTA can fly with either a sidestick controller or a center control stick, depending on the aircraft it is simulating. VISTA has been a stand-in for the F-22 because of its advanced mimicry. The specialized VISTA simulation computers are integrated with the standard production F-16 fly-by-wire computer control system.

The Multi Axis Thrust Vectoring (MATV) engine exhaust nozzle gives the VISTA F-16 a degree of controllability even in post-stall conditions when maneuvers relying on control surface deflections lose authority. In the best sense of a testbed aircraft, VISTA's MATV technology contributed to the use of such a nozzle on short takeoff and vertical landing (STOVL) variants of the F-35.

Testbed Douglas A-3 (BuNo 144867) stood in for the F-14 in radar and armament testing at Point Mugu, bailed first to Hughes and later to Raytheon. In 2012, this A-3 became part of the Pacific Aviation Museum Collection at Pearl Harbor. (U.S. Navy)

Following its use in VISTA and MATV work in the 1990s, the VISTA F-16 was assigned to the U.S. Air Force Test Pilot School at Edwards in October 2000.

Whale Mimics Tomcat

A Douglas A-3 Skywarrior twin-engine attack bomber, a type of aircraft affectionately nicknamed "Whale" due to its immense size for a carrier-based jet, became a testbed used by contractors as they developed F-14 Tomcat radar and armaments. First Hughes and then Raytheon operated this A-3. In 2012 it was flown to Naval Air Station North Island at San Diego and loaded aboard a Navy ship for transport to the Pacific Air Museum at Pearl Harbor, Hawaii.

Seaplane Slippers

The hydrodynamic aspects of seaplane hulls underwent years of study by industry and the NACA. Various hull shapes were devised to part the water effectively and safely. For the U.S. Navy, large seaplane operations were considered viable for two decades after the end of World War II.

Navy requirements included the ability to operate from rough water and to patrol at sea for hours. In an effort to optimize both aerodynamic and hydrodynamic efficiency, the NACA and the Navy collaborated on a unique hull testbed aircraft in 1947–1948. A Grumman J4F Widgeon amphibian (BuNo 32976) was modified with a radical hull

The modified Grumman Widgeon testbed nicknamed "Petulant Porpoise" gained a sparkling restoration and a place of honor at the huge Pima Air and Space Museum in Tucson, Arizona, as photographed in February 2016. A deep-chinned slipper nose attachment was one hull design tested on this aircraft. (Author)

Beefy flange and bolts allowed different seaplane hull slippers to be mated to the "Petulant Porpoise" testbed airframe. (Author)

Detachable slipper hull test designs (such as this) could be bolted in place on the Widgeon testbed. The Pima Air and Space Museum in Tucson, Arizona, displayed the aircraft and this second hull slipper in February 2016. (Author)

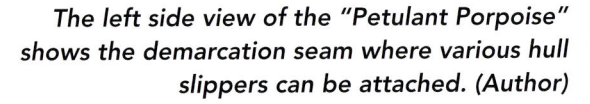

The left side view of the "Petulant Porpoise" shows the demarcation seam where various hull slippers can be attached. (Author)

that could have various nose and bottom shapes bolted on to test water-handling and flight characteristics. The resulting testbed aircraft received the nickname "Petulant Porpoise." Spanning 40 feet, it was a sub-scale surrogate for the Martin P5M Marlin flying boat, which had a wingspan of 118 feet 2 inches.

A horizontal flange ran from the tail to a spot about 2 feet ahead of the windscreen, where a vertical flange was added. These flanges allowed for mating with identical flanges on detachable hull bottoms and noses of different geometries for flight and water testing. The hull shapes were bolted at the flange lines, similar to the way in which some aircraft of the era used flanges to attach outer wing panels. Hull shapes included various locations for the characteristic notched step in the hull.

To accommodate the removable hull sections, the altered Widgeon's cockpit floor had to be raised and its controls modified.

The NACA Langley research facility had already made smaller models of a seaplane hull with a planing tail, a design that promised to diminish porpoising in the water but maintain low-water resistance and superior aerodynamic performance. The *Petulant Porpoise* took the planing tail out of the test tank and into open sea and sky on a larger scale. The NACA quantified the performance of various hull shapes and the Navy evaluated their utility to support its seaplane mission in the Cold War. This meant that the large P5M Marlin would be the chief beneficiary.

The *Petulant Porpoise*'s scaled P5M hull featured a high length-to-beam ratio, a shallow V-shaped step, and a long afterbody. Before the P5M Marlin's first flight on 30 May 1948, its new hull was estimated to shed 4 percent of the aerodynamic drag of the predecessor PBM Mariner hull while being 20 percent more efficient hydrodynamically.

After the Marlin hull test, a scale version of Convair's huge XP5Y-1 flying boat was scheduled. The NACA also designed a third hull add-on to explore its planing tail theories. The NACA design also promised greatly lowered aerodynamic drag in flight.

Research with the *Petulant Porpoise* was conducted at the NACA Langley Memorial Aeronautical Laboratory in Virginia, with nearby water access. Onboard instrumentation was developed by the NACA for the changeable amphibian. Shore-mounted high-speed cameras recorded water takeoffs and landings.

If seaplane advances were few in the post-war years, the Petulant Porpoise nonetheless stands as an innovative modular aircraft testbed. It survived, and as of this writing is displayed, along with a second hull slipper, in the Pima Air and Space Museum in Tucson, Arizona.

F-5 Attenuated Sonic Boom

In the centennial year of manned flight, NASA flew an ungainly modification of an F-5E (USAF 74-1519; BuNo 741519) to explore the ability to shape, and possibly attenuate, the intensity of sonic booms. The F-5E testbed featured an intentional deep-chinned forward fuselage that spoiled

With a deep chin that suppresses sonic booms while demolishing the usually pleasing aesthetics of the F-5E, this testbed for the SSBD was photographed over Palmdale, California, in August 2003. (NASA)

Chapter 11: Airframe Mods and System Tests

the F-5's legendary and lean shark-like form, but for good reasons. The modifications were executed by Northrop Grumman.

Flying over audio monitors on Edwards, the F-5E testbed was sometimes followed in flight by a stock F-5E to enable gathering of its baseline shockwave data in the same atmospheric conditions. The research flights showed a lessening of sonic boom severity with the whale-like modified F-5. The 2003 program was called Shaped Sonic Boom Demonstration (SSBD).

Scientists and engineers had predicted the ability to shape and minimize sonic booms since the early 1960s. Theories led to wind tunnel tests, but not until the F-5 SSBD was the concept put into full-scale actual test. In the 1990s some discussion was given to modifying an SR-71 Blackbird with the desired shape to mitigate sonic booms, but the cost and changes in research emphasis shelved that idea.[157]

The road to using a radically altered F-5E as the testbed began when Northrop Grumman received the go-ahead for a shaped sonic boom experiment in 2000 from the Defense Advanced Research Projects Agency (DARPA) under the Quiet Supersonic Platform (QSP) program. Northrop's first choice for the testbed vehicle was to use some available supersonic Teledyne-Ryan BQM-34E Firebee II drones then in Navy storage at Point Mugu, California. The Northrop Grumman team obtained the last of these supersonic drones known to be available.

The BQM-34E was initially attractive because of such things as low cost and modular construction that would make refitting new wings and shapes easy. The fact that the BQM-34E was a drone also meant that it did not need to have man-rated systems. Some previous wind tunnel studies dealing with modifying BQM-34s for supersonic research also made this selection look attractive initially; more detailed research suggested other parts of the BQM-34E's airframe could interfere with the work. And breathing life into these dormant surplus and stored vehicles was going to be no small task.[158]

Because the Firebee studies had shown the best approach for shaping a sonic boom involved reshaping and extending the forward fuselage, a Northrop Grumman aerodynamicist, David Graham, recalled the variations available in different noses for versions of the F-5E and F-5F Tiger II supersonic jet fighter. Although those variations did not prove to be useful, the Northrop Grumman team embraced the single-seat F-5E, a Northrop product with which they were familiar, as the testbed basis for their shaped sonic boom work.

The F-5's inlets were far enough to the rear to be out of the way; this looked like a good option for a testbed. In July 2001, DARPA gave the F-5E modification the go-ahead. Computer modeling, including computational fluid dynamics (CFD), aided in shaping the F-5E testbed's new forward fuselage. Northrop Grumman convened a Shaped Sonic Boom Demonstration Working Group with members from several aerospace contractors and research groups to refine the shape, leveraging CFD expertise from Boeing and Lockheed to complement CFD work conducted in house by Northrop Grumman. By now, the design was mature enough to warrant modifying Northrop Grumman's existing F-5E supersonic wind tunnel model. Tests at the NASA Glenn Research Center supersonic wind tunnel bolstered confidence even more.[159]

Northrop obtained the use of F-5E BuNo 741519 through a Cooperative Research and Development Agreement (CRADA) with the U.S. Navy. Northrop Grumman's El Segundo, California, facility designed and built the new forebody for the F-5E. It was installed at a Northrop Grumman depot facility that serviced F-5s at St. Augustine, Florida. New aluminum frames, bulkheads, and composite skin filled out the lower nose shape that gave this F-5E the look of a feeding whale more than the shark it started out to be.

The flight evaluations conducted at Edwards, where the NASA Dryden (later Armstrong) Flight Research Center is located, validated expectations and showed that a shaped sonic boom held its shape all the way to the ground. The potential exists for further development of shaped airframes to make supersonic overflights routine.

AWJSRA and QSRA

It has been argued that NASA is at its best when conducting pure aeronautical research that ultimately benefits the industry. Such is the case with a pair of modified de Havilland Aircraft of Canada C-8 Buffalo transports that explored increasingly complex forms of lift augmentation. The Buffalo, originally devised for the U.S. Army's fixed-wing transport needs, was never ordered in quantity in the United States once the Air Force assumed responsibility for such missions. The Canadian military and other operators bought the Buffalo, which remains a remarkable, short-field turboprop cargo aircraft.

Two demonstration C-8s originally bought by the U.S. Army as the C-7 served as testbeds for conversion during NASA's Augmentor Wing Jet STOL Research Aircraft (AWJSRA) and the even more ambitious Quiet Short-Haul Research Aircraft (QSRA). In both cases, Boeing executed the modifications.

The AWJSRA, based on C-8A 63-13686, featured a replacement wing of shorter span, mounting two Rolls-Royce Spey jet engines that provided traditional propulsion and augmentor airflow for the powered lift devices incorporated into the modified wing. As such, this modified Buffalo (sometimes referred to in America's Bicentennial year of 1976 as the "Bisontennial") can claim to be the world's first jet short takeoff and landing (STOL) transport demonstrator.

NASA's Quiet Short-Haul Research Aircraft (QSRA), a modified de Havilland C-8 Buffalo, had a new wing with four jet engines blowing exhaust over the upper surfaces of the wing, creating lift. The QSRA demonstrated a number of unarrested landings and free deck takeoffs from the aircraft carrier USS Kitty Hawk *(CV-63). (NASA Ames Research Center)*

Modified de Havilland of Canada C-8A Buffalo became NASA's Augmentor Wing Jet STOL Research Aircraft (AWJSRA) with a new wing of shorter span and new powerplants. A pair of Rolls Royce Spey jet engines replaced the original two turboprops of a traditional C-8 Buffalo. Boeing performed the modifications. Flights began in mid-1972. (NASA Ames Research Center)

Chapter 11: Airframe Mods and System Tests

The new short-span wing had augmentor flaps, blown ailerons, spoilers, and fixed leading-edge slats. Exhaust air could be passed between the upper and lower sections of the augmentor flaps to enhance lift. The airflow derived from the jet engines was cross-ducted to the augmentor flaps from each engine, giving a redundant capability to the flaps in the event that one engine was out. Strikingly, the jet exhaust from the two engines could be vectored from 6 degrees to 104 degrees below horizontal.

With characteristic meticulousness, NASA used wind tunnel tests at Ames to validate the augmentor flap design. It was possible to slow the modified Buffalo to 50 knots on approach to landing; 60 knots was routine for this, a telling success for the test devices fabricated into the modified de Havilland's wing. Takeoffs and landings over 50-foot obstacles could be achieved in less than 1,000 feet of horizontal travel.

After the essential powered-lift design features were tested, the AWJSRA received a digital guidance, control, and display system. This upgrade gave the aircraft computer control of pitch, roll, and yaw. It also enabled computerized thrust and thrust deflection operations. NASA has reported that work on the AWJSRA was useful when the U.S. Air Force was developing follow-on STOL transport specifications, ultimately yielding information valuable to the C-17 design. Among its accomplishments, the AWJSRA was the first to demonstrate fully automatic flight for a powered-lift STOL aircraft.

Although the AWJSRA had all the trappings of a NASA effort, the team at NASA's Ames Research Center partnered with Canadian counterparts on the project. Canada's Department of Industry, Trade, and Commerce joined the effort. Pilot Seth Grossmith from the Canadian Ministry of Transport joined NASA research pilot Bob Innis.

If the AWJSRA validated many concepts for jet STOL transports, a second ex–Air Force C-8A Buffalo (63-13687) pushed such developments to new heights. This Buffalo was the basis for NASA's Quiet Short-Haul Research Aircraft (QSRA). By the mid-1970s, jet transport noise around airports was an engineering problem. The QSRA project included the goal of producing STOL performance with the lowest noise levels that could reasonably be attained.

Boeing again received the contract to re-wing and modify the C-8A into the QSRA configuration. This time, the new wing featured modest sweepback. Most prominent was the placement of four turbofan engines atop and ahead of the wing to provide upper surface blowing. The chosen powerplants were Avco Lycoming YF102 high-bypass jet engines originally intended to power the Northrop A-9 attack jet, and previously tested in an engine testbed made from a North American AJ Savage bomber.

On the QSRA, the four engines' exhaust flowed over the wing's upper surface and flaps. This capitalized on the Coanda effect, with some of the propulsive force of the engines turned into propulsive lift. The first flight was in 1978. In stable flight conditions, the QSRA registered lift three times what would be expected on conventional aircraft.

Noise levels were measured at 90 EPNdB (Equivalent Perceived Noise decibels) at a specific 500-foot distance and position. This was the lowest measurement achieved at that time for a jet STOL transport. Testers said that the QSRA noise footprint was measurably smaller than that produced by a comparable conventional jet transport.

The QSRA went on to demonstrate fantastic STOL capabilities when it launched and landed on the aircraft carrier USS *Kitty Hawk* without using either catapult or arresting gear. Later in its program, the aircraft was fitted with a DFBW system and heads-up display (HUD). This led to HUD instrument approaches to touchdown, incorporating modifications in the control system to accommodate ground effect as the QSRA neared the runway surface.

QSRA results were added to the body of knowledge attained from earlier NASA-Ames STOL research that benefited Air Force and McDonnell Douglas C-17 program pilots who flew the QSRA to evaluate HUD and flightpath control augmentation. This influenced the C-17 design.

The bounty derived from the QSRA continued with its final research phase, called jump-strut tests. The hydraulics of the nose landing gear of the QSRA prompted nose-up rotation of the aircraft during its takeoff run, making takeoff distance even shorter.

NASA researchers found some specific uses of QSRA-demonstrated technologies to benefit other aircraft including the C-17. The agency also takes the long view in research programs such as these, confident that the knowledge gained now resides in the intellectual toolkits of future aeronautical designers.

Hiller X-18

The Hiller X-18 was an ambitious tilt-wing testbed design of the 1950s. George L. Bright and Bruce Jones, Hiller test pilots, made the first flight of the put-together Hiller X-18 (57-3078) on 24 November 1959 at Edwards. To prove the concept at lower cost, the fuselage of a Chase C-122 transport was adapted to become the X-18. An extremely low-aspect-ratio straight wing was fabricated with the ability to pivot from normal horizontal placement to vertical, with the leading edge and the two turboprop engines pointing upward.

A pair of scavenged Allison T40 turboprop engines originally intended for the proposed vertical-takeoff Lockheed XFV-1 and Convair XFY-1 experimental Navy fighters provided the primary power. T40s powered the contra-rotating Aeroproducts propellers that were variously reported as 16 or 14 feet in diameter. A Westinghouse J-34 jet engine mounted in the fuselage provided two-dimensional ducted exhaust for low-speed pitch control.

This pioneering attempt at a large tilt-wing aircraft provided useful research for later tilt-wing projects, including the XC-142. A potential problem in the X-18, later corrected in other tilt-wing and tilt-rotor designs, was the lack of interconnected power for the large propellers in the event of the loss of one engine. This would have rendered the X-18 uncontrollable. The frequent and gusty winds over the Mojave Desert at Edwards could also move the X-18 if the wind hit the broadside of the tilted wing. An Edwards AFB aircraft roster covering the period July to December 1960 lists the X-18's use as "VTOL ground effect simulation."

Following a harrowing propeller pitch issue that led to a spin in the X-18 in July 1961 on flight 20, the machine never flew again. By this time, it had been flown with wing tilting as high as 33 degrees. The X-18 subsequently was mounted to a special vertical-lift test stand at Edwards for testing until January 1964. Scrapping followed.

This short-winged X-18 shows the aircraft's unusual inflight configuration. Ducting extending to the rear gave two-dimensional pitch control at low speeds, with power from an internally mounted Westinghouse J-34 turbojet. After an abbreviated flight test program, the airframe faded from view, probably scrapped at Edwards after 1961. (AFFTC/HO)

The Hiller X-18 used the fuselage of a Chase YC-122C transport mated to new tilt wings with a pair of scavenged T40 turboprop engines. It flew, and also yielded data in ground tests, on a lift stand at Edwards. (AFFTC/HO)

Kaman K-16B

One hybrid testbed never flew before it was placed in a museum collection. The Kaman K-16B used the fuselage of an amphibious Grumman JRF-5 Goose (BuNo 04531) mated to a low-aspect-ratio tilt-wing that supported two GE T-58 turboprop engines turning large-diameter three-blade rotor/props.

The K-16B's wing, manufactured by helicopter maker Kaman, was a virtual parasol, riding above the fuselage of the Goose host. It could pivot 50 degrees, directing rotor wash and turbine exhaust downward and backward. This testbed retained Goose wing floats, with the addition of a linkage that allowed the floats to stay in their traditional horizontal position even as the wing pivoted upward. The wing spanned 34 feet, compared to 49 feet for a stock Goose.

The Kaman K-16B used the fuselage of an amphibious Grumman JRF-5 Goose (BuNo 04531) mated to a low-aspect-ratio tilt-wing that supported two GE T-58 turboprop engines turning large-diameter three-blade rotor/props. The testbed never flew, but it made full-scale wind tunnel tests. (NASA)

The Kaman K-16B during outdoor engine run is without its modified wing floats. The use of an existing Grumman Goose fuselage saved design and construction costs for this experimental testbed. (NASA)

It is said that the Navy wanted to explore the ability to use vertical and/or short takeoff and landing (V/STOL) amphibious aircraft to land in rough seas. The effect of downwash on this hybrid Goose near the ocean's surface, however, was never tested; the program that began when the testbed was finished late in 1959 was canceled in 1962 before any free flights were made. Some tethered and wind tunnel testing had taken place. The K-16B had a gross weight of 4 tons. Its speed was projected to be 200 mph.

As of this writing, the sole unflown Kaman K-16B resides in the New England Air Museum near Windsor Locks, Connecticut.

F-16XL Cranked Arrow

When an aircraft under flight test has finished its program, the prototype machines discarded by the military may wind up at one of NASA's flight research centers. Sometimes this only leads to the ultimate preservation of an oddball aircraft as a museum piece years later; at other times, it yields a viable testbed aircraft for further research.

General Dynamics developed a modified delta-wing version of its successful F-16 Fighting Falcon to compete with the F-15E for ground attack. That did not yield a production contract, but it did create two of the so-called cranked-arrow wing F-16 prototypes that found extended use at NASA's (then) Dryden Flight Research Center located on Edwards.

Years after the demise of America's original supersonic transport (SST) program in 1971, NASA engaged in research intended to foster a better SST. Under the banner of NASA's High-Speed Research Program, research applicable to a hypothetical High-Speed Civil Transport (HSCT) was begun. NASA probed a number of research avenues to see if practical solutions could be found for future supersonic aircraft sound levels and efficiencies. Part of this effort involved the two F-16XLs with modified testbed wing sections.[160]

The goal was supersonic laminar flow control (SLFC). The first phase, flown in 1991–1992, used the single-seat F-16XL (F-16XL-1) carrying a section of the upper left wing with tiny holes to promote suction as a means of SLFC. The most aggressive testbed configuration mounted a special wing glove to the left wing of the two-seat F-16XL variant. Its shape varied from the planform of the right wing, creating an unusual asymmetric appearance in a program identifying this aircraft as F-16XL-2. The glove had 10 million laser cut holes in it. It covered about three-fourths of the upper surface of the left wing, and 60 percent of the leading edge.

A follow-up to the successful, if operationally precarious, Air Force X-21A exploration, the NASA SLFC program boosted the research into supersonic speeds. Like the older X-21A, suction through the tiny holes was used to control boundary layer air over the modified wing. A flight test engineer in the back seat operated the suction device, tuning it incrementally in a quest for laminar flow changes.

The lopsided F-16XL flew 45 times between October 1995 and November 1996. The overall NASA High-Speed Research Program was canceled in 1999 before it could be completed.

A gloved wing for laminar flow studies made the F-16XL-2 planview asymmetrical, as seen in this 1996 view from its KC-135 tanker. Aerial refueling is a bonus for testbeds if a tanker is available, extending data-gathering flying time with fewer interruptions having to land and then take off again and climb back to test conditions. (NASA Photo by Jim Ross)

In 1996, F-16XL-1, the single-seat version, participated in the Cranked-Arrow Wing Aerodynamic Project (CAWAP) at Edwards, with tubing along the wing to obtain pressure distribution data. This supported NASA's High-Speed Civil Transport (HSCT) research. (NASA)

AFTI F-16

The Air Force's Advanced Fighter Technology Integration (AFTI) F-16 made more than 700 research flights in 10 trailblazing programs between 1978 and 2000. Flight control experiments, cockpit design, and targeting were among the contributions made by the AFTI airframe. Visible modifications included triangular chin canards that augmented maneuverability. Advances in autopilot and ground-collision avoidance were perfected and proven in the AFTI F-16. The overall goal of the AFTI F-16 program was to enhance capabilities for future fighter aircraft.

After developing an advanced digital flight control system, the AFTI F-16 effort moved to its second phase, the Automated Maneuvering Attack System (AMAS) in the early 1980s. This explored integration between the flight control system and the fire control system and avionics.

The AMAS effort of the 1980s was predicated on hostile environments with many targets and many threats. In that environment, the AFTI F-16 program sought to increase the likelihood of taking out a target on the first pass to minimize exposure. The AMAS exploration further presumed the need to use unguided munitions, increasing the workload on the AFTI F-16's flight control and fire control systems to

make ballistic computations allowing for weapons release and arrival on target while the aircraft was maneuvering. Goals included lateral toss bombing while maneuvering at 4 to 5 g at 200 feet above ground level.[161]

The need for ground-collision avoidance while engaging in radical evasive maneuvers and releasing weapons drove some AFTI/F-16 developments. A g-induced loss-of-consciousness (GLOC) recovery system developed in the AFTI/F-16 meant that a pilot could be physically incapacitated, yet the fighter could take action to avoid ground collision.[162]

The AFTI F-16's final test program contributed to F-35A development in 2000. In the following year, the special AFTI/F-16 was placed in the collection of the National Museum of the United States Air Force.

Quiet Spike

The ability to fly faster than sound is easier than the ability to fly quietly faster than sound. The shockwave associated with supersonic flight, although commonplace to residents around test areas such as Edwards, is often considered a nuisance to be avoided. Sonic booms generally relegated the Concorde supersonic transport to transoceanic flying when supersonic, and placed most populated landmasses off limits for commercial supersonic overflight.

In an effort to tame the effects of sonic booms, NASA worked with Gulfstream Aerospace beginning in August 2006 to test that company's patented Quiet Spike, a special non-concentric telescoping nose boom mounted on a NASA F-15B testbed aircraft. The boom, weighing about 470 pounds, was 14 feet long when retracted for subsonic flight, extending to 24 feet for supersonic flight. The purpose of the extension was to modify the shape of the sonic boom wave, in an effort to make it softer and quieter.

In the classic vein of aeronautical research conducted by NASA (and its predecessor the NACA), Quiet Spike research gives designers another tool, another channel to explore. The program ended in early 2007. Sonic boom modification was achieved, but the concept remained in the experimental stage.

The Quiet Spike F-15 nose boom is in the extended position, 25 September 2006. The boom has the ability to modify the characteristics of a sonic boom. (NASA Photo by Lori Losey)

Gulfstream's patented Quiet Spike extendable nose boom was mounted to a NASA F-15B for tests over the Mojave Desert in 2006–2007. (NASA)

Chapter 11: Airframe Mods and System Tests   157

Chapter 12

MISCELLANEA

The modified Gulfstream II became a trimotor with the addition of a special propfan engine on the left wing. A long rod on the wing held microphones to capture propfan noise levels; a shorter boom on the wing was a dynamic boom with flutter accelerometers that helped guard against flutter issues with the modified aircraft. (NASA)

Some aircraft defy clear categorization as testbeds, motherships, or parasites, but they still deserve recognition. Other ideas embraced the mothership rationale but never advanced beyond the research and feasibility stages.

Jenny Mounts a Glider

Around World War I the Air Service devised a way to carry target gliders atop the center section of Curtiss JN-6 Hispano-engine Jenny trainers, to be released in flight. Another early NACA photo depicts a JN-4H Jenny with a test wing shape suspended on cables beneath the biplane.

Hydro-Ski Tests the Water

Traditional seaplanes and floatplanes rely on some form of watertight buoyancy to remain afloat on water. Hydro-skis are planing devices that keep an aircraft out of the water only so long as sufficient forward speed is maintained. With loss of speed, the hydro-ski aircraft must either settle in the water, or, if that is not desirable, it must beach itself or otherwise secure a location above water level.

Convair explored hydro-skis for the radical delta-wing XF2Y-1 Sea Dart seaplane in the early 1950s. At rest, the Sea Dart floated on its hull, the hydro-skis below water. Other

Early in the realm of motherships, some U.S. Curtiss Jennies were modified with a wing mount plus a tail support for carrying and launching a GL-1 target glider. This example was redesignated JN-6HG-1. (1FW/HO)

landplanes and amphibians became testbeds for hydro-skis in the 1940s and 1950s. The first known hydro-ski application was conceived in 1947 and it was discovered that hydro-skis tended to perform better in rough seas than did traditional seaplane hulls.[163]

Traditional seaplane hulls were not meant for rough-water open-ocean operations beyond the level that their structure could tolerate. Making them larger was not the answer; if open-ocean rough-sea flying boats were to succeed, a starkly different approach was needed. The genesis of hydro-ski design originated with conceptual jet designs, but the rough-water capabilities of skis gave the idea another rationale.

The NACA put twin hydro-skis on a tow-tank model of the D-558 research aircraft in 1947, learning much in the process. Meanwhile, the Edo Corporation, long known for its float designs, was exploring an Air Force request for a jet fighter design capable of operating on snow, ice, and water, and intended for arctic operations. A strengthened keel was key to this design proposal until the NACA research became known, at which time hydro-ski design gained prominence with the Edo team.

It was a classic example of the way in which the NACA benefited the industry by researching concepts for the further exploitation by companies such as Edo. Back at the NACA tow tank, a 1/8-scale model of a Grumman OA-9 Goose hulled amphibian modified with hydro-skis showed promise. By October 1948, Edo flew a full-size OA-9 with a hydro-ski modification. The Air Force's unusually specific request for a specialized arctic jet fighter design was not pursued by the service.[164] Nevertheless, the U.S. Navy, impressed with the modified OA-9, stepped in with funding to pursue hydro-skis for water-based airplanes.

Edo modified a second Grumman Goose, this time a Navy JRF-5 version, for the Navy, incorporating a single-ski concept. Navy interest included a novel landplane design intended for Marine Corps use, in which an aircraft in motion could use the water for takeoff and landing, but it could start and stop from a beach or special barge to keep it safe when there was insufficient speed for the hydro-skis to be effective. All-American Engineering was involved in this early concept.

By 1952, a special ski-equipped SNJ-5C Texan trainer was flying, and a hydro-ski Cessna OE-1 joined the effort the following year. The single-ski JRF-5 configuration showed promise, although some transitional lateral instability might cause problems for some pilots, so twin ski designs were pursued for better stability on the water.

The next hydro-ski aircraft was a JRF-5 Goose twin-ski modification tested in 1953. The twin skis could contain integral beaching wheels to make the design even more useful.[165]

The non-seaplane hydro-ski testbeds by now had included a classic Piper Cub, a Stinson OY (the Navy/Marine version of the L-5) an XL-17D Navion, and an Army de Havilland U-1A Otter. It was the early 1950s, and Martin was grooming its big, new P5M Marlin twin-engine flying boat, recently joining the U.S. Navy as a patrol plane. The Marlin enjoyed some post-war hull design refinements, but it was still a standard seaplane with a hull.

Edo ambitiously suggested to the Navy that a 25-ton Marlin should be fitted with hydro-skis. The Navy went a bit more conservatively on the suggestion. Instead of modifying a new Marlin, the service contracted with Martin and Edo to modify an older (but still large) PBM Mariner seaplane with a strut-mounted hydro-ski. The test ski could be set at three different strut extensions on the modified PBM-5. As with smaller aircraft, the big PBM showed improved rough-water

Chapter 12: Miscellanea 159

handling but still maintained marginal lateral control when the single ski emerged. The hydro-ski PBM Mariner flew from 1955 to 1958. Two sizes of single hydro-ski were tried on the PBM, and the later, smaller, version showed better lateral stability. A fire in its hangar destroyed the Mariner before all testing could be accomplished.

Into the mid-1960s, a Lake LA-4 amphibian was modified with the addition of a hydro-ski for further testing by Thurston Aircraft.[166] Other aircraft were considered for hydro-skis, including an unbuilt variant of the C-130 Hercules four-engine transport.

A different concept, vertical floats for seaplanes, used two non-flying PBM-5 Mariners in 1963 to compare how steady they floated in the ocean; one was a stock example and the other PBM used vertical floats that extended into the water and raised the Mariner out of the water. The vertical float version made the Mariner much more stable. A report on the project said: "The addition of vertical floats to the PBM seaplane eliminated motion sickness and discomfort of the crew while resting on the water for extended periods in Sea States I and II."

Although these were not the most violent of seas, the stability imparted by the vertical floats prompted the testers to advise continuing the research in heavier seas as an aid to ASW (anti-submarine warfare) aircraft. (The P-5 Marlin seaplane was still in U.S. Navy service at the time, and the potential for a new ASW seaplane existed.) It was suggested that production vertical floats could be readily stowed on a seaplane.[167]

Combination Night Fighter Notion Not Developed

A British writer came up with the notion of a combination night fighter for the light wing loading of a mothership plus the speed and agility of a pursuit plane to give the night fighter pursuit greater range and endurance. (He was no doubt inspired by his nation's historic interest in combination aircraft such as the World War I Bristol Scout fighter plus Porte seaplane combination or the 1930s Maia/Mercury piggyback.) The result, christened the Slip-Wing Fighter, graced the cover of the May 1941 issue of the American version of *Flight* magazine, an aviation periodical that used, during that era, boldly colorful renderings of often-fanciful aircraft on its cover.[168]

Ultimately, twin-engine night fighters similar to the American Northrop P-61 Black Widow solved the range and speed issues to a reasonably satisfactory degree without resorting to hitching a ride. The fascination with combination aircraft still surfaced in the popular press from time to time.

Soviet Nuclear Mothership Concept

In the competitive Cold War year of 1956, *American Aviation* magazine devoted space to a piece of Soviet artwork purporting to show a giant notional nuclear-powered mothership that could fly in the stratosphere at speeds ranging from 6,000 to 12,000 mph. The mothership could potentially stay aloft on its nuclear power for weeks, with smaller craft landing on it dorsally to discharge passengers, cargo, and mail to be delivered to an earth-bound destination by another aircraft connected to this high-speed atmospheric orbital system. It is interesting to note that the 1956 crystal ball that predicted nuclear-powered aircraft did not, evidently, acknowledge the way in which the unheard-of Internet would make mail a very different priority.[169]

Getting the Jump on Ejection Seats

Several Air Force aircraft never intended to receive ejection seats nonetheless became ejection-seat testbeds. The XP-47N airframe, without an engine, was fitted with an ejection seat for testing and demonstrating at Wright Field after the war. A B-29 at Muroc Army Air Base in 1947 hosted ejection seats in the tail and bomb bay according to the base historical record for that year. A Northrop P-61 Black Widow (42-39498) also made a feasible ejection-seat testbed with its spacious fuselage and widely spaced twin vertical tails that provided a clear shot upward and aft. Nicknamed "Jack-In-The-Box," this P-61 hosted the first in-flight ejection by an American in 1946. First Sgt. Lawrence Lambert made the successful ejection test.

The U.S. Navy tested an early Martin-Baker ejection seat in the modified gunner's area aft of the wing of a Douglas JD-1, the Navy's variant of the A-26 Invader. On 30 October 1946, Lt. (jg) A. J. Furtek first ejected from the JD-1 testbed high over the landing circle for blimps at Naval Air Station Lakehurst, New Jersey, clearing the single vertical fin and

First Sgt. Lawrence Lambert was assisted into the modified gunner's compartment on a Northrop P-61 Black Widow called "Jack-In-The-Box" before the first inflight ejection by an American in 1946. The light-colored stripes are typical on testbed aircraft to provide orientation to the angle of attack, useful in tracking and measuring the path of anything or anyone separating from the aircraft in flight. (Air Force via Dave Menard Collection)

The Douglas JD-1 Invader used by the Navy as an ejection-seat testbed enabled ejections to be entirely separate from the cockpit environment. The red bar in the national insignia and the desert locale indicate that this undated photo was not from the original Navy ejection tests over New Jersey in 1946. (U.S. Navy/NARA)

rudder of the Invader. Lieutenant Furtek's trailblazing ejection followed unmanned ballasted tests of the seat earlier.

Furtek's early seat employed the Navy's preferred method of seat actuation, which is pulling a stiff fabric cover mounted to the seat downward and over his face, activating the rapid ejection sequence while providing a measure of protection from wind blast.

At that time in Navy ejection seat exploration, the service was looking into borrowing an F-82 Twin Mustang for more ejections at higher speeds.[170]

Bell P-83 and Ramjets

Only two Bell P-83 jet fighters were built. They were too late for World War II and too slow for the jet age. Conceived as a long-range escort fighter in the early days of jet propulsion when efficiency was poor and aerial refueling nonexistent, the big P-83 relied on the ability to tank a lot of fuel to obtain the needed range. When no production orders were forthcoming, the first of the two XP-83s was modified with the addition of two ramjet engines, one under each wing. Conceptually, the P-83 might be powered at altitude by the two ramjets alone, and not its regular J33 turbojets.

On the first flight test of the ramjet installations in September 1946, however, the ramjet under the P-83's right wing began to burn. As the heat and fire attacked the wing structure, pilot Chalmers "Slick" Goodlin lost use of the right aileron. Goodlin and test engineer Charles Fay, riding in a special compartment fitted inside the XP-83's large fuselage, bailed out and the burning XP-83 crashed into the dense forests of upstate New York, its testbed career cut very short.

Chapter 12: Miscellanea 161

The Bell XP-83 was a big experimental fighter that did not see mass production. One of the two built was fitted with a ramjet under each wing (shown) for tests to see if the ramjets alone could power the P-83 once it attained sufficient speed from its normal pair of wing root–mounted J33 turbojet engines. A fire in one of the ramjets put an end to the program when it spread to the wing of the XP-83, forcing pilot Chalmers Goodlin and a test engineer to bail out. (AFFTC/HO)

The NACA harnessed this Northrop F-15 Reporter reconnaissance aircraft (in an era when "F" meant Foto, not Fighter). The high-altitude capabilities of the Northrop F-15 suited it for releasing aerodynamic models from altitudes as high as 43,000 feet over Muroc (later Edwards). The released shape was used for obtaining transonic speed data as it plunged earthward. Technicians oscillated the remotely controlled and instrumented shapes through differing angles of attack before decelerating them with air brakes and recovering them with parachutes. (Garry Pape Collection)

NACA Motherships with Drop Models

Muroc Air Force Base historical records for 1948 indicate that a Grumman F7F Tigercat, probably part of the NACA's Ames laboratory in the San Francisco Bay area, flew sorties at Muroc to drop aerodynamic shapes. The NACA used a number of airframes for this kind of aerodynamics research. One of the few Northrop F-15 Reporter photo reconnaissance offshoots of the P-61 Black Widow night fighter was used by the NACA Ames research facility between February 1948 and October 1954. This aircraft, serial number 45-59300, carried NACA number 111. The NACA also borrowed a P-61C 43-8330 (later designated

162   Testbeds, Motherships & Parasites

This gritty black-and-white image shows a NACA Black Widow carrying a centerline-mounted ramjet test vehicle high enough to be in contrail-producing conditions. The P-61 is the C-model (43-8330), part of the Smithsonian Institution Collection. (NASA Ames Research Center)

ERF-61C) that belonged to the Smithsonian Institution, for use as a drop vehicle at Edwards. The high-altitude capabilities of the F-15 Reporter suited it for releasing aerodynamic models from altitudes as high as 43,000 feet.

Weighted for acceleration, the instrumented models passed through the transonic speed region, an area of intense interest at that time. The models were controllable to achieve different angles of attack. After gathering data, speed brakes and parachutes slowed the models for safe recovery. The development of transonic wind tunnels eventually supplanted this type of testing, but in its day it was a creative use of motherships.

After the NACA finished with the F-15 Reporter it was sold as surplus, passing through several owners who made modifications. Ultimately flying as a firebomber in the 1960s, this same aircraft came to grief in a takeoff mishap in California in 1968. The P-61C was returned to the Smithsonian, where it remained in storage for decades until going on display in the National Air and Space Museum's Udvar-Hazy Center near Dulles International Airport.

The NACA, like its successor, NASA, enjoyed reasonable access to a variety of military airframes that could serve the agency's aeronautical research goals. The NACA was studying the effects of shockwaves around ramjet inlets at supersonic speeds. In 1948, a North American F-82 Twin Mustang (originally the last of two XP-82 prototypes, 44-83887) carried a tapered 16-inch tubular ramjet with tail fins, releasing it from an altitude of 30,000 feet. This enabled the ramjet to reach speeds of Mach 2.4 in descent, telemetering data to a ground station by way of a spike antenna in the nose of the ramjet. Before the Twin Mustang took the ramjet aloft, it had been released from a B-29.

NACA Ramjet Testbeds

One of the most laborious and alternatingly frustrating but satisfying aerospace endeavors since World War II has been the investigation of ramjets as simple, fast powerplants. Unlike the pulse jet engine of Germany's wartime V-1 buzz bomb, a ramjet requires forward motion of sufficient speed to ram air through an inlet to a compression chamber where the introduction of gas vapor and flame can cause a continuous combustion that produces thrust.

Some successes, such as the Boeing Bomarc with a Marquardt ramjet, stand out, while the general state of the art has been a work in progress for more than seven decades as this is written. Scramjets (supersonic combustion ramjets) were validated with NASA's X-43, which achieved at least Mach 9.65 on 16 November 2004. However, in the heady post-war 1940s, it seemed to some publicists and futurists as if high-speed ramjet-powered aircraft were just around the corner.

To explore the parameters of ramjet technology, the NACA's Flight Propulsion Laboratory in Cleveland, Ohio, flew an Air Force B-29 Superfortress (45-21808) mounting a long ramjet tube beneath the bomb bay, in the manner also used for conventional jet engine testing. It has been reported that this B-29 also served as a jet engine testbed.

Ramjet tests began in April 1947. The ramjet was 52 inches beneath the B-29's fuselage in test conditions. Tests were conducted between altitudes of 5,000 and 30,000 feet and at speeds up to .51 Mach. The ramjet's angle of attack could be adjusted independently of the B-29. A periscope protruding from the B-29 afforded testers a view of the ramjet in operation.

These test flights helped quantify performance of a ramjet of the era. Ignition could be achieved via spark up to an altitude of 14,000 feet. When flares were used for ignition, the useful altitude was boosted to 30,000 feet. Combustion efficiencies, response time to change in fuel flow, and minimum fuel-air ratios were mapped.

This NACA B-29 ramjet testbed was photographed in 1947 with its bomb bay doors open, revealing the curved cutout to surround the streamlined mount for the ramjet when the ramjet was extended in flight and the doors were closed around it. The open-air fire truck is also an asset of the NACA. (NASA Glenn Research Center photo via Dennis Jenkins)

The NACA used B-29 45-21808 as a ramjet testbed in the skies over the Midwest when this photo was taken in 1948. The small, dark extension beneath the lower nose may be the periscope that was fitted to the B-29 to enable inflight observation of the ramjet. The ramjet's angle of attack could be set independently of the B-29. (NASA Glenn Research Center)

This glossy-black Northrop P-61B (42-39754) was one of several Black Widows used by the NACA as testbeds in the immediate post-war years. This example featured a ventral extension beneath the yellow buzz number (PK-754) that probably carried cameras to record activity of a ramjet typically carried beneath the belly of this P-61. (NASA Glenn Research Center)

The NACA's P-61B (42-39754) carried a fixed wide horizontal inlet ramjet for tests when photographed with the ramjet operating in 1947. This ramjet shape was envisioned for applications within airfoil sections. (NASA Glenn Research Center Photo by Eldon Holst)

The second XF-82 Twin Mustang (44-83887) was used by the NACA (also known under NACA number 132). It carried a 16-inch ramjet for a free-flight launch at Wallops Island, Virginia. The F-82 also appears to be toting three 100-pound practice bombs, possibly for ballast. (NACA/NARA)

Bearing a three-letter alphanumeric buzz number, XF-82 44-83887 sustained this incident on 14 December 1949 at the NACA Cleveland, Ohio, facility. (NACA/NASA)

Ground crew members stand ready to pull the chocks so that a NACA F-82 Twin Mustang mothership can take off with a ramjet to drop, circa April 1949 at Cleveland, Ohio. (NACA Photo by Bill Wynne)

North American F-82B 44-65168 (NACA 132) carries a test ramjet beneath the center wing section. This F-82 is the Air Force's Betty Jo *record-setting Twin Mustang subsequently placed in the National Museum of the United States Air Force. The ramjet is a 16-inch model with an internal rocket booster. The device is about to be dropped over the NACA's Wallops test range in 1952. (NARA/NACA)*

The XF-82 ran into trouble in mud in December 1949 at Cleveland. (NACA)

The NACA Twin Mustang has a ramjet test drop rocket, circa 1953. The F-82B Twin Mustang 44-65168 was the Air Force's Betty Jo, *a record-setter that flew from Hawaii to New York on 27–28 February 1947, covering 5,051 miles. It was the longest non-stop flight ever made by a propeller-driven fighter. In 1957,* Betty Jo *returned to the Air Force for display. (NACA via Dennis Jenkins)*

The F-82B 44-65168 flew for a time, circa 1952, in NACA markings as NACA-132. Here, the aircraft is carrying a ramjet tube. (NACA via Dennis Jenkins)

The NACA also operated Northrop P-61B Black Widow 42-39754 in 1947 as a host aircraft for a ramjet with a widened horizontal inlet and configuration. The testbed was flown from the agency's Aircraft Engine Research Laboratory in Cleveland, Ohio. (A name change in April of that year made this the NACA Flight Propulsion Research Laboratory; in September 1948 it became the NACA Lewis Flight Propulsion Laboratory; with the creation of NASA it became the NASA Lewis Research Center in October 1958; on 1 March 1999 it was renamed the NASA John H. Glenn Research Center at Lewis Field.)

A cutout in the center section of an F-82 wing between the two fuselages accommodated center mounting of ramjet drop rockets, as seen in this 1951 photo. (NASA Glenn Research Center)

The flattened ramjet configuration was envisioned for employment in confined aerodynamic spaces such as a wing or tail airfoil. With a standard service ceiling above 33,000 feet, the Black Widow could take the ramjet to high altitudes to evaluate the experimental powerplant's combustion efficiency, thrust, flame stability, and starting characteristics in the low-pressure low-temperature environment found at altitude.[171]

Marquardt's Ramjets

Roy Marquardt was a Northrop engineer during World War II. Studying engine airflow issues, he embarked on a quest to develop useful ramjets. A ramjet literally rams air through its inlet to compress it before combusting a fuel-air mixture to produce thrust. Because of this, a ramjet must be moving before it can operate.

Marquardt founded Marquardt Aircraft and provided a pair of ramjets to the U.S. Army Air Forces to be mounted to the wingtips of P-51D Mustang 44-63528 in 1946. In this way, the Mustang was a testbed with the indispensable task of getting the ramjets moving through the air fast enough to sustain combustion. The Navy also experimented with Marquardt ramjets on an F7F Tigercat, and the NACA delved into their use.

On 21 November 1947, the first flight of an aircraft powered solely by ramjets was claimed when Lockheed test pilot Tony Levier throttled back the J33 jet engine of a special F-80A Shooting Star (44-85214) so that its airspeed was sustained by a pair of wingtip-mounted Marquardt ramjets only.

Marquardt Aviation in Venice, California, built ramjets 20 inches in diameter and 7 feet long, as well as ramjets 30 inches in diameter and 10 feet long for Air Force research. The larger ramjets were mated to the F-80.

By 1948, the F-80 testbed had made more than 100 flights with ramjet engines. The F-80 first attained a speed of 400 mph at 20,000 feet, using its standard jet engine. Igniting the Marquardt engines on the F-80 initially created a spectacular flame exhaust plume as long as 40 feet behind the wingtips. As engine operation began to settle in, the flames diminished to about 15 to 20 feet, and when the ramjets were in their full stride, the flames disappeared.[172]

Marquardt's pioneering ramjet efforts eventually went beyond add-on test engines on aircraft. Boeing IM-99 Bomarc surface-to-air missiles used Marquardt ramjets after achieving sufficient velocity on vertical launch from a rocket engine. A similar Marquardt design powered the D-21 reconnaissance vehicle mounted atop modified M-21 variants of the Lockheed A-12 Blackbird.

Marquardt Gorgon Navy Missile

In the gray area between munition and vehicle, the Navy's Gorgon IV missile was carried aloft and launched from a Navy-operated Northrop P-61C (43-8336) Black Widow night fighter borrowed from the Army Air Forces. The Gorgon, built by Martin, might have become a guided air-to-surface missile but was used as a radio-controlled research vehicle to validate its ramjet propulsion. To balance the weight and drag of the 22-foot-long Gorgon under the right wing of the Black Widow, a drop tank was carried under the left wing for ballast. Test launches in 1948 over the Point Mugu range on the Pacific Ocean gave way to follow-on work at Chincoteague, Virginia, which was closer to the Martin plant in Maryland.

The late 1947 first flight of the Gorgon IV was claimed as the first flight of a ramjet-powered vehicle in the United States. Other ramjet tests before that had involved free-falling or test-stand variants, not flying airframes. Gorgon IV carried 116 gallons of gasoline, said to be sufficient for a 10-minute powered flight. The Gorgon was launched by the P-61 at a speed of .5 Mach; it is necessary to have forward speed to push air into the ramjet to make it function, unlike a pulse jet that can operate while stationary. Cruise speed for the Gorgon was set before launch, within the capabilities of the ramjet, which was rated to provide up to .85 Mach. To adhere to the programmed cruise speed, fuselage-mounted speed brakes automatically extended and retracted as needed in flight.[173]

Lincoln Turboprop Tests

Photos from 1947 show a British Avro Lincoln four-engine bomber with standard piston engines on the inboard mounts and a pair of turboprop Bristol Theseus engines in the outboard positions. This aircraft was probably Lincoln RA716/G. Sometimes the testbed Lincoln was flown with the two Merlin piston engines shut down. This testbed also flew with Rolls-Royce Avon turbojets in the outboard positions. The Lincoln found favor as a British engine testbed. Built too late for World War II combat, the Lincoln was a large, low-time airframe available in the post-war years.

In 1948, a still-camouflaged Avro Lincoln (probably RF530) served as the testbed host for a Napier Naiad 1,500-hp propjet mounted in the bomber's nose. It was a utilitarian nose job that placed the streamlined silver Naiad nacelle at the apex of a very blunt nose cap on the Lincoln. This same Lincoln, in civil registration, flew circa 1956 with a Rolls-Royce Tyne propjet in the blunt nose.

Armstrong-Siddeley Python turboprops and Bristol Proteus turboprops graced the outboard engine mounts of Lincolns after the war and another nose mount hosted a Napier Nomad diesel turbo-compound engine. A Bristol Phoebus turbojet occupied the bomb bay of a Lincoln, and an afterburner-equipped Rolls-Royce Derwent jet engine was tested aboard a Lincoln in a ventral mount location. The Phoebus, a test engine not intended for production, helped prove components for the Proteus turboprop.

An Avro Lancastrian (VH742) flew with two Rolls-Royce Nene jet engines on two nacelles and two regular Merlin powerplants, circa 1946–1948. This is said to be the first jet airliner (a bit of a stretch) whenever this testbed carried non-crewmembers on demonstration flights where the two piston Merlin engines would be shut down. Lancastrian airframes, derived from the famed Lancaster bomber for transport use, saw extensive service in the United Kingdom as engine testbeds after the war. Lancastrians also flew with two Avon, Ghost, and Sapphire jet engines for test purposes. As post-war developments of the V-12 Merlin and Griffon piston engines continued, testbed Lancastrians carried a pair of Griffon 57s in addition to the usual Merlins, a pair of Merlin 600s along with standard Merlins to test the Griffons for the Avro Shackleton patrol bomber, and Merlin 600s for the Canadair C4M and Avro Tudor transports.

It was either a Lancaster or a Lincoln with three-blade propellers that hosted an Armstrong-Siddeley Mamba propjet in its nose in 1947, with the engine exhausting ventrally from an aft-facing tube below the testbed bomber's cockpit.

French Spoils of War

France was in the unusual position at the end of World War II of possessing virtually no usable air force, yet having access to much German equipment, some of it built in occupied France. A quick way to jump-start post-war France's air force was to revive German bombers until new aircraft were available.

The French aircraft industry possessed two unfinished airframes for the German Heinkel He 274, a high-altitude four-engine strategic bomber that did not fly in German hands. Heinkel was constructing the He 274s at Suresnes, France. Following the liberation, they were completed under the French designation AAS-1. The French Air Force used them initially for high-altitude research before French industry used them as motherships, carrying test airframes aloft on dorsal struts.

The AAS motherships were replaced with the indigenous four-engine SNCASE SE.161 SNA Languedoc airliner design as a mothership later in the 1950s, for air launch tasks for French high-speed aerodynamic research prototypes.

168   Testbeds, Motherships & Parasites

Czech, Please

In 1970, an Ilyushin Il-14 tricycle-gear twin-engine transport in Czechoslovakian markings mounted a Czech domestic turboprop engine, the Motorlet M-601, in a chin position. Looking every bit the Soviet-era no-nonsense industrial installation, the IL-14 testbed vouches for the universality of using a larger airframe to safely flight test a new powerplant. The Motorlet M-601 engine initially developed 540 hp and was forecast to evolve into a propjet of 740 hp.

Supersonic Trimotor F-106

As part of the United States' technical inquiry into the feasibility of a supersonic transport, on 20 October 1966 NASA's Lewis Research Center (later renamed Glenn Research Center) took delivery of an Air Force F-106B Delta Dart supersonic interceptor. The goal was to mount a pair of J85 jet engines, one under each wing, to test supersonic inlet and nozzle designs. This effectively made a trimotor of the supersonic Delta Dart. To chase this fast testbed, NASA also obtained an F-102A Delta Dagger interceptor.

NASA technicians removed the F-106's weapons system and 700 pounds of wiring, according to NASA Glenn Research Center. The NASA team cut holes in the F-106's wings and modified the trailing-edge elevons to accommodate the two J85 jet pods. The underwing mounting of the J85 nacelles replicated the general notional layout of the larger engines envisioned for the SST. The former missile bay of the F-106 hosted a 228-gallon fuel tank installed by NASA; some original fuel-tank capacity now carried test instrumentation.

The J85s were operated on F-106 flights at speeds between Mach 1 and Mach 1.5. Under one wing the J85 carried a nozzle for which performance characteristics were known.

NASA 616, a borrowed USAF Convair F-106B, banked away from the camera plane in this 1969 photo, revealing the two J85 jet pods under the wing that made this testbed a trimotor F-106. The goal was to test supersonic inlet and nozzle designs. (NASA Glenn Research Center; Photo by J. David Clinton)

The NASA F-106 testbed at Cleveland underwent servicing of one of its underwing J85 engines. (NASA Photo by John Marton)

Chapter 12: Miscellanea 169

Six supersonic nozzle variations as well as two inlet designs were tested on the other J85 engine. The designs had already been wind tunnel tested, but shockwaves inside the confines of the tunnel limited these evaluations, making free-air testing necessary.

The proximity of Lake Erie provided an uninhabited supersonic corridor 200 miles long for the NASA delta-wing trimotor. A 1,100-mph dash could be made at 30,000 feet in minutes. The rapid consumption of fuel at these speeds meant the F-106 sometimes returned to Lewis Research Center low on fuel. For this reason, flights were made under VFR conditions to remove the additional hurdles of instrument flying while low on fuel. When a different high-performance jet crashed at Lewis in July 1969, the F-106 testbed program was moved to Selfridge Air Force Base in Michigan. NASA pilots commuted from Lewis to Selfridge where they made the supersonic runs.

By May 1971, Congress had voted to stop funding the government's support of SST research. Between February and July 1971, the NASA F-106B executed acoustic and aerodynamic tests of various jet engine ejector nozzles. This included subsonic testing at .4 Mach.

For these slower speed tests, the F-106's main engine was throttled back, and the Delta Dart flew at 300 feet over a recording site, with additional microphones placed to the side of the path of flight. A conical exhaust nozzle developed by NASA Lewis was the baseline device. Nozzles with 32 or 64 radiating spokes showed some promise in flight. However, "In the end, no general conclusions could be applied to all the nozzles," according to a brief NASA Glenn Research Center explanation about the project.

The NASA F-106 trimotor tested this cooled plug nozzle, a GE 32-spoke nozzle, in 1971. (NASA Photo by Paul Riedel)

This is the left-hand engine mounted under wing of NASA F-106B, photographed in 1968. Ingenuity was required each time NASA needed to modify an existing aircraft for a purpose never conceived by its original designers. (NASA Photo by Martin Brown)

From X-Plane to Testbed

If the experimental North American Aviation X-15 rocket aircraft resides outside the bounds of this book, the stretched X-15A-2 version's final flights gets in just under the wire. On what turned out to be the last three flights of the number-2 X-15, this modified rocket plane was a flying testbed for the aerodynamic shape of a supersonic ramjet under investigation by NASA. The end result of a multi-million-dollar rebuild of this X-15 was to make it a supersonic ramjet testbed. The fuselage was extended more than 2 feet to accommodate the ventrally mounted ramjet with its propellant tankage inside the fuselage addition.

The three-engine F-106 testbed flown by NASA to assist with SST nacelle design was photographed in 1968. (NASA Photo by Martin Brown)

Bold NASA markings adorn the F-106B three-engine testbed in this photo, circa 1976, showing two different nozzle configurations on the add-on J85 engine pods. The dark stripe handily, if a bit awkwardly, obliterates USAF markings, including the serial number on the vertical fin. (NASA Photo by John Marton)

At Cleveland in the summer of 1972, NASA's three-engine F-106B testbed parked beside an F-102 used as a chase aircraft. (NASA Photo by J. David Clinton)

Chapter 12: Miscellanea

On 3 October 1967, this X-15, carrying two huge, external, propellant tanks for a longer rocket burn to generate faster speed, was released from its B-52 mothership. A special ablative coating covered the X-15 in an effort to protect its metal skin from friction-caused heat damage. Even the cockpit windows were shielded and altered in geometry for this mission. Mounted to the ventral fin of this modified X-15 was the shape of a representative supersonic combustion ramjet with which NASA and the Air Force intended to gain data on its high-speed aerodynamics before attempting to fly an operable ramjet.

The ultimate goal was to have an externally mounted ramjet, in addition to the X-15's regular XLR99 rocket engine, for flight testing of the supersonic ramjet at speeds between Mach 3 and Mach 8. Notional artwork from 1964 depicts a flattened ramjet design similar to the much later X-43. The shape attached to the X-15 for flight testing was circular in cross-section with a central spike in the inlet.

The mission profile took the X-15A-2 to a record speed of Mach 6.7, or 4,520 mph. Supersonic shockwave impingement on the test ramjet shape mounted to the X-15's modified ventral fin concentrated a jet of heat that overwhelmed the ablative coating. About 25 seconds after X-15 engine shutdown, the superheated shockwave burned through the skin of the ramjet pylon. The X-15 was damaged as heat migrated internally. Air Force Major William J. "Pete" Knight, a skilled pilot, landed the stricken X-15A-2 safely on Rogers Dry Lake.

Even as plans were under way to repair the X-15A-2, funding for the ramjet test program was cut and the damaged X-15 was repaired for display at the U.S. Air Force Museum at Wright-Patterson.[174]

Johnny Armstrong, an aerospace research engineer assigned to the X-15 program, wrote an eloquent technical report about the X-15A-2 testbed program in 1969 that was cleared for public release decades later. Armstrong's text brings the program to life as he explains how decisions were reached and what happened on the fateful flight. Armstrong's descriptions are interesting and informative; his official report has been excerpted as Appendix 4.

Blackbird Booster

Some of Kelly Johnson's celebrated Blackbird triple-sonic spy planes were groomed to be motherships for drones. Prior to launch, the Mach 3 D-21 rode dorsally atop a modified A-12 Blackbird. The D-21 carried a camera in a jettisonable package. After completing a photo reconnaissance sortie, the low-observable D-21 jettisoned the camera where it could be retrieved. The D-21 then self-destructed. Problems plagued the D-21 program, including one post-launch collision with the mothership M-21 version of the A-12 resulting in the death of one crewman on 30 July 1966. D-21s shifted to using B-52H motherships with underwing pylons after this. Subsequent tests, as well as unsuccessful real-world missions over China between 1969 and 1971, failed to fully validate the concept.

Improved satellite imagery as well as international politics contributed to obsoleting the D-21 program. Years later, unflown D-21s were made available to several museums and displays in the United States.

C-130 Balloon Launcher

A practical reversal of John J. Montgomery's pioneering exploits took place at the Air Force Flight Test Center between February and November 1977. Where Montgomery used a balloon to get his airframes aloft in the early 1900s, the Flight Test Center evaluated using an airplane to get balloons aloft.

The evaluation report summarized the effort: "This test program was conducted to investigate the feasibility of developing an air-launched balloon system, which is eventually intended to carry a tactical communications relay to altitude when a ground launch of the system is not practical. Fourteen airdrop tests were conducted from a C-130 aircraft at altitudes of 10,000 feet and 25,000 feet MSL and at an indicated airspeed of 130 knots. The test system weighed approximately 1,500 pounds and operated in three stages: (1) extraction and descent under the extraction parachute, (2) deployment and descent under the recovery parachute, and (3) deployment and filling of the balloon.

"The test results indicated that the system developed is a feasible approach for air-launching a balloon system with some further development. When testing was terminated by the customer due to a shortage of balloons, time, and money, several problems still existed in the system. The first-stage suspension system components were susceptible to damage during the extraction phase of the test, and the balloon inflation tube was susceptible to being choked off and bursting during the balloon-filling phase."[175]

The Air Force Geophysics Laboratory requested the test, which had an Air Force acronym, ALBS, for Air-Launched Balloon System. The intent was to create a quick-reaction lighter-than-air tactical communications relay platform that could be positioned where needed: "Operational planning calls for the system module to be extracted from a C-130 aircraft at 25,000 feet. When the system is properly deployed in midair, the stored ALBS balloon will be extended vertically and filled from a cryogenic helium storage unit. The balloon will then carry the communications relay to its assigned altitude (70,000 feet) while the inflation hardware floats to the ground by parachute."[176]

Fourteen airdrops were made from a C-130. The gross weight of the units dropped ranged from 1,473 to 1,570 pounds. A 28-foot ringslot drogue chute pulled the unit from the C-130. A 42-foot Pioneer parachute supported the device after extraction. Two high-pressure helium bottles inflated the 157,000-cubic-foot balloon; they were capable of giving the balloon 4 percent inflation. That amount of inflation, and references in the report to the "dummy balloon," suggests the tests did not deploy a full balloon capable of ascending, as would an operational device.

The timing and mechanics of the deployment were described in detail: "At drop initiation, the 28-foot chute deployment bag was released from the C-130 aircraft pendulum mechanism. The drogue line and then the 28-foot chute were pulled out of the deployment bag as it fell away from the aircraft and the 28-foot chute inflated and pulled the box out of the aircraft (time, 0 seconds).

"At T = 10 seconds, the box and extraction subsystem had stabilized in a vertical descent . . . under the 28-foot chute and Radioplane #1 [release] fired, releasing first-stage suspension slings from the box. The box fell away and the main parachute deployment line pulled the 42-foot chute and the donut pack, attached to the apex of the 42-foot chute, out of the box and the 42-foot chute began inflation.

"At T = 14 seconds, two reefing line cutters fired, allowing the snood to open, releasing the excess canopy slack it had contained. The canopy of the 42-foot chute then completed inflation.

"At T = 16 seconds, four reefing line cutters fired, cutting the lacing holding the cover onto the donut pack.

"At T = 20 seconds, Radioplane #2 fired, releasing its attachment to the 42-foot chute apex through the main parachute deployment line, and the balloon deployment line pulled the balloon out of the donut pack.

"At T = 35 seconds, a valve in the helium subsystem opened and allowed helium gas to pass up the balloon inflation tube and into the balloon."[177]

The chain of events required for a full deployment of a balloon sometimes led to malfunctions in the test drops, but the conceptual reversal of John J. Montgomery's idea remains intriguing as a means to deploy balloons with specialized packages where needed. A T-28 chase plane from the National Parachute Test Range (NPTR) paced the C-130.

Trimotor Beech 18 for Continental

Continental Motors, makers of general-aviation opposed-cylinder engines, used a smaller testbed aircraft to advantage. Starting with a twin-engine Beech Model 18 (N7927C) that a previous owner had converted to use Continental engines in the standard wing mounting locations, Continental commissioned Rawdon Brothers Aircraft in Wichita, Kansas, to fair a test engine and cowling neatly into the nose of the former Twin Beech. The result was a very aesthetic small trimotor useful to Continental for engine testing purposes. The basic Beech had been built as a C-45G (51-11785) for the U.S. Air Force. The airframe was scrapped in 2001.

Unducted Fan Testbeds

Everything old is new again; the unducted fan concept tested by Boeing on a 727 (N32720) in 1986 used contra-rotating exposed and enlarged fans made by General Electric.

McDonnell Douglas continued the exploration on the prototype DC-9 Super 80 jetliner N980DC (later re-designated as an MD-80/81) in 1987. Called ultra-high-bypass (UHB) by Douglas and unducted fan (UDF) by Boeing and General Electric, the exposed fans were said to give jetliner speeds that were "quieter and more fuel-efficient than today's best turbojets," according to a McDonnell Douglas spokesman.

The UHB engine was mounted on the left side only of the MD-81 twinjet testbed. The ultra-high-bypass testbed

Exposed contra-rotating fan blades as part of a GE powerplant occupied the right engine mount on this Boeing 727 testbed in 1986. (General Electric)

McDonnell Douglas was confident that its MD-80 testbed that carried one ultra-high-bypass GE fan engine would segue into the MD-91 commercial jetliner based on the MD-80/DC-9 series. This did not happen, and the MD-91 went unbuilt, making the stretched MD-90 with twin V2500 turbofans the final member of the long-lived DC-9-80 series. (McDonnell Douglas)

flew over Edwards in the summer of 1987 in the new McDonnell Douglas livery. That year the company confidently predicted the UHB technology would begin airline service in 1991 on the advanced MD-91 airliner.[178]

McDonnell Douglas perceived a niche for propfan engines in the mid-1980s. The DC-9/MD-80 series of airliners were in production, and the aft-fuselage-mounted engines lent themselves to easy conversion to propfans, something that would be more difficult or impossible to do with the wing-mounted engines on the airliners of competing companies. Therefore, McDonnell Douglas promoted the value of UHB engines based on fuel efficiency.

The first iteration that the company tested was the GE GE36 engine. Two different sets of prop blades were tested, the second set proved to be quieter than the first. Next, in 1989, the testbed airliner hosted a 578-DX propfan developed by Allison and Pratt & Whitney. The MD-91 and other derivatives were not made. An easing of a fuel crisis may have cooled any airline interest in the new type of powerplant. It has been suggested that the appearance of visible, large-diameter fans looked like a throwback, and the selling of the idea may have run into perception issues.[179]

Interestingly, the McDonnell Douglas MD-81 testbed started life as a traditional Douglas DC-9-81 with conventional JT8D fanjet engines. This 1979 aircraft was involved in a test flight incident at Edwards in May 1980 when the empennage broke free after a hard landing during FAA certification testing for high-sink-rate landings over a 50-foot obstacle.

NASA Propfan

In the thick of all this propfan testing in the latter half of the 1980s, NASA and various industry partners collaborated on a propfan testbed. The basis was a Gulfstream II business jet twin. It retained both aft-mounted jet engines and acquired a purpose-built nacelle for a propfan installation on the left wing. Called the Propfan Test Assessment (PTA) program, flight testing was conducted at Dobbins Air Force Base, near Marietta, Georgia. That is where the modified aircraft's prime contractor, Lockheed-Georgia, is located. Other industry contributors to the program included Lockheed-California, Rohr, Gulfstream Aerospace, Hamilton Standard, and Allison. Gulfstream performed modifications to the aircraft.[180]

A propfan with broad blades and a diameter of 9 feet was a major element of the NASA-sponsored Advance Turboprop Program (ATP). This program included the PTA program contract with Lockheed-Georgia using the Gulfstream, the Large-Scale Advanced Propeller (LAP) contract with Hamilton Standard, the Unducted Fan (UDF) contract with General Electric, and the Advanced Gearbox Technology contracts with Pratt & Whitney and Allison.

The PTA effort had two main objectives: propfan structural integrity and noise reduction. The propfan on the Gulfstream wing was powered by a combination of an Allison model 570 industrial gas turbine engine and a T56 gearbox. Modifications were completed in time to begin static testing at Rohr's test site at California's Brown Field, near San Diego. Flight testing on the Gulfstream began in March 1987, backed by wind tunnel testing to check performance, flutter, stability-and-control, and handling characteristics. Flight testing concluded in March 1988 with the evaluation of a lightweight, acoustically treated cabin interior.[181]

Modification of the Gulfstream II necessitated changes to the fuel and hydraulic systems as well as other systems to accommodate the third engine mounted to the left wing. Of particular significance was the addition of a static balance boom on the opposite wing weighing more than 1 ton to help offset the 6,500-pound engine system on the left wing. A dynamic boom was added to the left wing for flutter safety. The wing structure was beefed up with doublers in strategic locations and a new rib near the nacelle attachment point. During early flight testing, 700 pounds of 3/8-inch stainless steel was installed inside the cabin to protect the crew in the event of a prop blade failure.[182]

Tests called for evaluation of several parameters, including forward speed, altitude, and inflow angle of attack. The angle-of-attack changes were not achievable within the nominal center-of-gravity changes that occurred with fuel burnoff. Lockheed devised a tiltable nacelle to achieve the needed inflow angles of attack. By changing the forward-to-aft nacelle mounts, the tilt angle could be adjusted from 2 degrees to 3 degrees downward, yielding the desired inflow conditions over a range of flight operating scenarios. The propfan engine was not operated during takeoff because of structural restrictions on flap hinge moments. The Allison propfan engine was started at an altitude of about 5,000 feet, using bleed air from the Spey jet engines at the rear of the Gulfstream.[183]

The testbed yielded data on propfan performance and safety as well as sound levels. The surface of the Gulfstream's fuselage was instrumented with 44 microphones concentrated near the plane of rotation for the propfan. This allowed mapping of propfan noise contours in the areas of likely maximum levels. A long, acoustic boom mounted on the left wing of the Gulfstream held five microphones at distances similar to the fuselage-mounted microphones. One of the reasons for mirroring the fuselage microphone locations was to determine the acoustic impact of the direction of rotation of the propfan. The propfan blades on the inboard side rotated upward; on the outboard side they rotated downward, and likely created different acoustic patterns with airflow.

Some noise testing was conducted at NASA's Wallops Flight Facility in Virginia, with the propfan testbed overflying ground microphones. High-altitude noise sampling was based out of Dobbins, flying into northern Alabama.[184]

C-5 Launches a Minuteman

Although the C-5A Galaxy was not destined to become the Space Shuttles' mothership, a C-5 made mothership history on 24 October 1974. That day a 43-ton Minuteman missile successfully fired its engine and climbed above the Pacific Ocean after being rolled out the back of the C-5 in flight in a feasibility test for air-launching intercontinental ballistic missiles.

The difficulty of making such air launches operational was deemed to be greater than the benefit, although it is said that the proven air-launch capability was a point of negotiation in U.S. strategic arms talks with the Soviet Union.

After a long career as an airlifter, the missile-launching C-5 (69-0014) was retired to the Air Mobility Command Museum at Dover Air Force Base, Delaware.

747 Tanker Dry Run

The prototype Boeing 747 was a stand-in for several projects, including feasibility studies of using this huge airframe as an Air Force tanker. Boeing was awarded a contract for $1.62 million in early 1972 to use the 747 in this effort. Under consideration by the Air Force Systems Command's Aeronautical Systems Division was a multipoint refueling system that used a flying boom under the aft fuselage plus two more refueling opportunities, one at each wingtip.[185]

The Boeing 747 prototype with an aerial refueling boom conducts a dry hookup with an SR-71 during a tanker test program at Edwards, circa 1972. (Boeing Photo by Vern Manion)

A B-52G is dwarfed by the Boeing 747 prototype as the two aircraft perform a dry hookup to evaluate the 747's viability as a tanker, circa 1972. (Boeing Photo by Vern Manion)

The most spectacular portion of the study was a series of dry hookups between the 747, carrying a refueling boom, and a series of Air Force aircraft, including SR-71, B-52, FB-111, and F-4. The 747 tanker concept was not pursued by the U.S. Air Force.

Micro-Fighters and Motherships

The concept of carrying fighters to the fray attached or inside a large mothership has resurfaced several times in the evolution of aviation. Decades after the U.S. Air Force's experiments with projects such as the XF-85, Tom-Tom, and FICON, a study conducted in 1972 and 1973 considered the use of so-called micro-fighters, which could be transported in a large aircraft such as a C-5 or 747, and be capable of airborne launch and recovery.

The U.S. Air Force's Flight Dynamics Laboratory at Wright-Patterson, then part of the Air Force Systems Command, sponsored the research contract for the study conducted by Boeing's Tactical Combat Aircraft Project.[186]

The study, Investigation of a Micro-Fighter/Airborne Aircraft Carrier Concept, suggested the viability of micro-fighters and flying aircraft carriers: "This concept feasibility study has provided the initial step toward development of an advanced concept of operation: the Micro-fighter/Airborne Aircraft Carrier. The operational employment of strike fighters operating from airborne aircraft carriers is indicated by this study to be technically feasible. Furthermore, the system concept offers the potential of great national benefit in a political world that leans toward a low-profile American exposure overseas while being responsive to diverse needs of our allies."

The study acknowledged the evolution away from parasite fighters in the 1950s that was made possible by advances in tankers, among other things. New technologies, however, might make the concept embraced by the old XF-85 Goblin feasible: "The modern concept for airborne launch and recovery combines new transport technology and emerging fighter technology to produce a system concept that goes beyond in-flight refueling to add in-flight rearming and multi-sortie capability for each fighter." The study proposed a 747 variant that would have employed air locks and pressure bulkheads to allow for two launch-and-recovery bays for small jet fighters to be stowed internally.

The micro-fighter summary said: "This investigation has studied the feasibility and usefulness of an airborne airbase and has found it to be technically feasible and potentially valuable to the nation as a rapid deployment multi-purpose strike system. It has the potential for intercontinental response, with large combat forces, before an aggressor can fully mobilize for invasion of neighboring countries."

The study had three goals: Evaluate the feasibility of the micro-fighter/airborne aircraft carrier concept; develop a micro-fighter point design that would enable a quantity of such fighters to be carried, launched, and recovered by a mothership of the 747 and C-5 class; and build a wind tunnel model of such a fighter. The 1973 study had to forecast when such micro-fighter and airborne aircraft carrier combinations might be constructed. It said that the initial concept would be based on 1975 technologies, using much off-the-shelf equipment. Planners optimistically, and perhaps realistically, expected capabilities to be even better for the micro-fighters by 1980. Potential growth in 747 weight capacity was considered in the study.[187]

The micro-fighter, like the XF-85 before it, had constraints on its size and geometry imposed by the need for internal stowage aboard the airborne aircraft carrier. A 747 could carry as many as 14 micro-fighters. Their wingspan was limited to 17 feet 6 inches. The fuselage length of all micro-fighters considered in this study was uniform at 30 feet. The height of the diminutive fighters ranged from 5.85 to 6.86 feet, thanks in part to the lack of traditional landing gear and the lack of need for rotating the fuselage on takeoff or flaring for landing, since these would be accomplished as releases and captures in flight (although emergency landing on skids was part of the formula). With internal armament and internal fuel, the micro-fighter launch weight was intended to be 10,000 pounds. The study did note that, "An overload capability of 40 percent was determined practical for air-to-ground applications."

An estimate of weight to be carried by the mothership was made: "Fighters, fuel, and air-to-ground weapons for three sorties each represent a carrier expendable load of approximately 200,000 pounds." Various basing scenarios included deploying the carriers with micro-fighters from the United States, the United Kingdom, and possibly Diego Garcia in the Indian Ocean. In an era dominated by the potential for conflict by peer air forces, the micro-fighter study said: "High-intensity combat against many types of Soviet aircraft would require air superiority roles for the fighter both as CAP (Combat Air Patrol) for interdiction missions and fleet air defense."

When mid-1970s use of micro-fighters over the Middle East was contemplated, the study noted: "Self-defense capability should include maneuver performance equal to MIG-21 without salvo of external stores. This requirement was found to be very sensitive to MF [micro-fighter] wing loading and thrust to weight ratio."

Five micro-fighter design layouts were considered. The intent was for each to be powered by a single YJ101-GE-100 jet engine. A delta-wing design had two vertical tails mounted just inboard from the wingtips. This version included both overwing and underwing mounting of missile weapons. A variable-sweep version was proposed, as were a canard-equipped variant, an arrow-wing design, and one design burdened with the acronym VITAC (Variable Incidence and Transisting Aerodynamic Center). The notion of a variable-sweep micro-fighter might seem logical to accommodate confined space aboard the mothership, but in fact the swept-back aspect was useful in high-speed flight, and yet a variable-sweep fighter had to be airworthy while swept back at slower launch speeds. The report summary noted: "Vehicle designs that employ folding or sweeping surfaces must be flyable at launch and recovery speeds in folded configuration."[188]

The envisioned micro-fighters would save weight, space, and complexity by deleting traditional landing gear. But they would have an emergency shock-absorbing skid system and a drag chute in the event return to the carrier aircraft was not possible. After studying several emergency-landing options, including inflatable cushions, a simple metal skid prevailed because of its low cost, light weight, and volume. Basic armament, in addition to releasable stores such as AIM-9E missiles, was to include two

Chapter 12: Miscellanea   177

M-39 20mm cannons with 400 rounds of ammunition. All five notional micro-fighters were premised to carry 2,500 pounds of jet fuel.

The two best micro-jet configurations for intended mission performance were judged to be the delta wing and the tailless variable-sweep design. The delta was simpler to build; the variable-sweep version had an overall performance advantage. A merging of some delta and arrow design features dictated by wind tunnel testing made the final cut.

When carrier mothership design tenets were considered, the important factors included the use of dual launch/recovery bays in the fuselage, inflight refueling booms, and the ability of the carrier aircraft to operate in other roles, as troop or cargo carrier or tanker. Any perceived benefit of the C-5's existing rear-loading doors was false. The summary noted: "The C-5A aft location for cargo off-loading is not usable with fighter-size vehicles (without extensive modification to airframe and flight control system). Bomb bay type arrangements close to the carrier center of gravity allow launch-and-recovery operations for vehicles up to 15,000 pounds."

The Boeing 747 was forecast to have greater growth potential for weight carrying than the C-5A; the 747 had better endurance and speed than the Lockheed C-5, according to the study summary. The 747 was also deemed to be easier to configure with dual refueling probes, one for each launch-and-recovery bay; ultimately the report favored using the 747F as the baseline for further studies of the carrier aircraft concept.

Rationale for the recovery of micro-fighters began with the refueling hookup. During about 30 seconds of contact with the refueling boom, the micro-fighter would be fully fueled. The boom would then begin to retract, pulling the fighter with it to reach the trapeze locks. Once the fighter was locked to the trapeze, power would be supplied to the fighter and the refueling boom would be stowed out of the way. Engine shutdown on the fighter would initiate the hoisting action of the trapeze to bring the micro-fighter aboard. An overhead crane would move the stowed fighters for maximum storage space.

Much like naval aircraft carrier crews train for rapid re-arming and turnaround of the ship's complement of aircraft, the crew of the flying carrier aircraft would be poised to re-arm the newly fueled micro-fighters to enable quick follow-on sorties. Trolleys would bring munitions to the aircraft as the fighters hung from overhead mounts in the fuselage of the 747F. Turnaround time per micro-fighter was estimated at 10 minutes.[189]

It was estimated that 44 people would be aboard the airborne aircraft carrier. This included 14 micro-fighter pilots, 18 support specialists, and 12 crewmembers for the 747. The summary noted: "Operation of 10 fighters in combat situations from a high-altitude base requires pressurized crew compartments and hangar decks. The launch-and-recovery bays become airlocks to transfer the fighters between environmental extremes."

Planners envisioned that airborne aircraft carriers would work with radar assets, including airborne AWACS and ground radars to find hostile targets. The safe defeat of inbound Mach 3 aircraft demanded launch of the first micro-fighter 90 seconds after detection of the enemy. The report claimed: "From an alert status [pilot in cockpit] two MF interceptors could be launched in approximately 80 seconds." Subsequent fighter elements could be launched at 80-second intervals.

To accomplish this with the dual-bay 747, an airlock pressure manifold would cycle pressurized air between the two bays as needed. Payload capability of the carrier aircraft was sufficient to enable three sorties per micro-fighter, replenishing armament and fuel from supplies aboard the 747F.

With so many potentially hostile air forces of the 1970s operating variants of the MiG-21, it must have seemed only logical to premise micro-fighter performance on the ability to thwart MiG-21s. The micro-fighters were designed with performance that would, on paper, be able to out-accelerate a MiG-21 without resorting to jettisoning external stores.

The micro-fighter study looked ahead to 1980s evolutions that might compete with jets having performance similar to the F-16 of that era, and predicted a stowable jet fighter with inflight thrust reversing capability that would enhance maneuvering. Direct side-force control was envisioned to enable flat turns for precision maneuvers. An inclined pilot seat and high-g cockpit were planned.

The advanced micro-fighter design was given the nomenclature Model 985-121; the earlier arrow-wing design was the Model 985-20. The ongoing development of the micro-fighter design included the need to reduce weight because the carrier aircraft reached gross weight before its volume was physically filled with fighters and supplies. The study said that this called for the development of a lightweight 25mm cannon with caseless ammunition and a low-cost defense missile. For the follow-on micro-fighter, the expectation was that one of these missiles would be carried in each wing, in internal tubes faired with frangible leading-edge covers. A finless design for the missile was posited, using vectored rocket thrust. Smart bombs and Sparrow missiles were said to complement the armaments available to the micro-fighter.

The study summarized differences between the 1975 and newer iterations of the micro-fighter: "The 1980 version of the Arrow MF is similar to the 1975 version but includes some configuration and structural changes that result in slightly different aerodynamic characteristics and considerably less weight. The changes that influence the aerodynamics were primarily the thicker wing-root sections, the reduced volume and shortened fuselage, and the internal carriage of the two air defense missiles replacing the external AIM-9s of the 1975 version. At most operating regimes these changes tended to favor the -121."

The advanced version of the micro-fighter was expected to be capable of air-to-air combat at altitudes up to 50,000 feet and speeds up to Mach 2.0. Launch for air-intercept missions was envisioned at 30,000 feet at .8 Mach. Trade-offs were acknowledged: The proposed internal missile stowage aided performance and external mounting of AIM-7 missiles posed a drag penalty that might be minimized if folding-fin versions could be developed to allow cleaner mounting to the micro-fighter.[190]

The study went into detail about possible air-to-air encounters. The micro-fighter would fly part of the sortie at subsonic speeds, part at supersonic or transonic speeds. Assuming a mission radius of about 350 miles from the carrier 747F, it was expected that the 985-121 fighter would have sufficient fuel for 10 360-degree turns at maximum thrust in the transonic speeds considered typical for air-to-air encounters.

The team studying micro-fighters said that deployment would be served best by air-refuelable 747s working with air-refuelable 747-based AWACS aircraft. The proposed 747 AWACS was supposed to have capabilities beyond the operational 707-based AWACS that would make it a better match for the 747 airborne carrier aircraft. The 747 AWAC/C, as it was called, was an Airborne Warning And Control/Command aircraft. It was intended to have its own two launch-and-recovery bays for a pair of reconnaissance micro-fighters. (It is perhaps appropriate to remember that this paper study was made at the height of the Cold War when defense budgets were not as restricted as they later became.)

Proponents forecast an ambitious Initial Operational Capability (IOC) for the micro-fighters by 1980 that would rely on a Multi-Purpose Strike System (MPSS) with 10 747 airborne aircraft carriers (AAC) and 1 747 AWACS, plus 100 micro-fighters. This roughly equated to four squadrons. The promise was deployment to Europe in eight hours from time of alert, on station and combat-ready. The method of carrying the fighters overseas in motherships allowed the fighter crews the luxury of being rested en route, instead of flying their own jets.

In this scenario, 20 percent of the fighters were to be configured for fleet air defense against all threats up to Mach 3.0, with the remaining 80 percent of the micro-fighters available for CAP, reconnaissance, reconnaissance/strike, or Close Air Support (CAS). Presuming an expendable load of 200,000 pounds, the 1980-version AAC was expected to be able to remain on station for 4.5 hours at a 2,600-nautical-mile radius while each of the fighters flew three or four sorties from 100 to 250 nautical miles in radius.

The study also suggested that global deployment could be accomplished with the use of yet another 747 variant, a tanker version. With such an armada, it was expected that global coverage from the United States could be achieved in 17 hours with one refueling en route to the farthest air-launch position envisioned. A total crew complement of 525 was figured for such a Multi-Purpose Strike System.

Planners said that the MPSS was an aerial version of the U.S. Navy's seaborne strike force, "self-contained completely for the duration of operations away from its home base." Interestingly, the study suggested that micro-fighters in a carrier aircraft offered an alternative to V/STOL fighters. One of the desired traits for the AAC motherships was "performance versatility for launch and recovery in clear air without contrails."

The micro-fighter study team no doubt knew the sometimes-troubled history of other attempts at launching and recovering fighters from bombers. The study recommended that wind tunnel testing of the micro-fighter should include initial testing to "measure interference effects at each event during the recovery."[191]

Ultimately, traditional means of force projection prevailed and the micro-fighter concept remained a paper airplane.

The Case for Engine Testbeds

The Society of Experimental Test Pilots (SETP) dwells in the rarified atmosphere of professional test pilots, people who must match seat-of-the-pants skills with scholarship and courage with common sense. In that atmosphere, in 1971 Rolls-Royce chief test pilot James Harry Pollitt offered his observations on the necessity of using flying testbeds for aeronautical engine development.

Pollitt argues that while bench testing at low altitude can start to validate an engine design, only a flying testbed can confirm the engine's performance where forward motion is involved, where altitude is an issue, where inlet design can prove to be critical, where things such as temperature, pressure, and humidity changes exist, and where dynamic changes in the condition of an aircraft and its engines can be introduced into the test mix.

Pollitt says that some flying testbeds of the past were less than optimal if they failed to introduce a realistic inlet for the jet engine under test. Another pitfall can happen if the testbed aircraft cannot replicate at least a meaningful portion of the engine's anticipated speed range. Single-engine flying testbeds are limiting.[192]

Pollitt describes a number of British engine testbeds. A twin-engine Canberra bomber hosted a new nacelle with the Viper turbojet beneath the right wing. This gave powerplant redundancy. Pollitt says, "The fact that the test installation was not a prime mover is considered a valuable asset."

When the Olympus 593 jet that would power the Concorde supersonic transport needed to be flight tested, a Vulcan bomber hosted a realistic engine installation ventrally, delivering high fidelity in the performance data.

At the time he presented his paper to SETP, Pollitt was anticipating good results from another realistic engine and airframe representation mounted to the belly of Vulcan XA903 to test the RB199 engine for the then-unflown MRCA (Multi Role Combat Aircraft), known better as the Tornado. Pollitt described the test installation: "The installation has a representative half-fuselage and single-engine installation. The test engine is not a prime mover, which allows the engine to be flown at a very early stage of development."[193] An early trial of a new engine has long been a recognized benefit of employing a flying testbed with redundant powerplants providing a safety margin in case the new engine suffers problems in flight.

Pollitt looked back at British flying testbed efforts and created a list of desirable traits when developing an engine for a particular aircraft:

At the Engine "Flight Prototype" Stage
- A complete engine/intake installation should be as near as possible to the proposed aircraft.
- The engine should not be a prime mover, which will allow a flight clearance to be obtained at a much earlier stage.
- A large aircraft is preferable as it allows space for instrumentation and multi-crew. The larger aircraft also permits stable aircraft conditions while severe handling on the test engine is being carried out.

When the Engine is Approaching Type Certification
- The prototype aircraft should be modified to give a production standard of engine and installation.
- Wherever possible this aircraft must be solely for engine development.[194]

When comparing the benefits of a flying testbed to the use of a ground-bound test facility, Pollitt described a scenario that could place a worthy demand on engineers in a test program: "There is a discipline that only flight test can impose upon engineers. They are acutely aware that they are dealing with human lives and, as they are an extremely responsible group of people, it makes them examine all their decisions in minute detail. 'Nearly good enough' is probably acceptable for the [fixed test facility] but is not in the ball park for a flying mission."[195]

The arguments for using multi-engine testbed aircraft abound; the examples, ranging from five-engine B-17s to B-52s and 747s, prove the point.

EPILOGUE

This review of testbeds and motherships can only whet the appetite for even more. Any trip through the holdings of technical aviation magazines of any era is likely to reveal yet another aircraft turned into a testbed or mothership in the endless quest to keep the state of the art sharp and moving forward.

Although this book captures the machines involved in so many testbed and mothership ventures, it is humbling to realize that these aircraft were only possible because of the brilliance and vision of engineers and builders matched with the skill and bravery of resourceful test pilots.

The award for longevity must go to NB-52B 52-0008, with more than four decades of mothership and testbed service. For creativity in motherships, it's hard to beat the clean-sheet designs emanating from Scaled Composites. As long as an aircraft's onboard fuel capacity is finite, as long as safety demands the use of unpiloted test vehicles, as long as economy dictates the modification of one aircraft to learn something new about aeronautics, the world of motherships and testbeds will remain vigorous and adaptive.

APPENDIX 1
TESTBED AIRCRAFT

Aircraft Type	Serial/Registration	User	Purpose	Timeframe	Comments
AJ Savage	N68667	Avco Lycoming	Jet engine testbed	1970s	BuNo 130418
A-20G	NX63148	Hughes Aircraft	XF-11 vertical tail		
JD-1 (A-26) Invader		USN	Ejection testbed	1946	
A-26 (B-26) Invader	43-22494	USAF	XQ-1 target carrier	Circa 1951	
A-26 Invader		General Electric	Radar-autopilot-armament system	Circa 1953	Boom above nose
A-26B Invader	44-34137	USAF	Tire research	Post-WWII	Clipped-wing
A-3 Skywarrior	BuNo 144867	Hughes/Raytheon	F-14 armament	1970s–2000s	
A-30 Baltimore	BuNo 09804	Martin	Transonic tests	Circa 1946–1947	
B-17G/299Z	44-85734/N5111N	Pratt & Whitney	5-engine testbed		
B-17G/PB-1	44-85747/BuNo 83999	Allison	5-engine testbed		XT-40, T-56
B-17G	44-85784	USAF/General Electric	Wingtip pod		
B-17G/299Z	44-85813/N6694C	Curtiss-Wright	5-engine testbed		
B-17G/PB-1	44-85571/BuNo 83992	CALSPAN	5-engine testbed	1946–1948	Not actually used
B-23		Lab for Electronics		Circa 1967	Near Boston, Massachussetts
B-23	39-050	General Electric	Turbosupercharger	1942	
B-24C	40-2385	General Electric	B-29 gun system	November 1942	Ground accident 1943
B-24C	40-2386	General Electric	B-29 gun system	Last B-24C	
B-24E	42-7057	General Electric	B-29 gun system		
B-24H (XB-24J-FO)	42-95100	General Electric		1944–1950	I-40 Testbed
B-24J	42-73215	Convair/General Electric	J-35 jet for XB-46	Circa 1946	Single-tail B-24 mod
B-24L	44-49916	General Electric	XB-47 tail gun test	Circa 1946	
B-24M	44-41986	NACA	Jet engine ice testing		Also windscreen icing
XB-24Q	44-49916	Emerson, GE	B-47 tail guns	1946	Armament and jet tests
B-25		Hamilton Standard	Propeller tests	Circa 1940s	
B-25	May be N39E	General Electric	FC5 flight control	Circa 1950s	
B-26/JM-1 Marauder	BuNo 66599	Martin	19B turbojet testbed	Circa 1946	Westinghouse engine
B-26G Marauder	43-34584	Snecma	Atar tubojet testbed		French testbed
XB-26H	44-68221	Martin	Landing gear test	1946–1947	Stump Jumper
XB-29	41-002	Boeing	Company testbed	Circa 1942–1947	
B-29	42-6357	NACA	B-29 engine tests	Circa 1944	
B-29	42-24716	General Electric	B-29 gun system	August 1944	
B-29	42-63571	General Electric	XB-36 tail gun		Returned to AAF
B-29	Unknown	USAF	Ejection seat tests	1947	Muroc (Edwards)
XB-29G	44-84043	General Electric	Bomb bay jet test	1945–1953	Former B-29B
B-29	45-21791	Pratt & Whitney	J42 testbed	Circa 1947	
B-29	45-21808	NACA	Ramjet testbed	Circa 1946–1948	
P2B-2S (B-29)	84030/45-21791	USN	Presumed anti-submarine warfare tests	Circa 1947	
P2B-2S (B-29)	84031/44-87766	USN	Presumed anti-submarine warfare tests	Circa 1947	
B-45		Pratt & Whitney	J57 testbed		
B-45		Pratt & Whitney	J75 testbed		
B-45A	47-049	Westinghouse	Engine testbed		
EB-45A	47-096	Curtiss-Wright	J67 Engine testbed	Circa 1954–1955	Final B-45A built
EB-45C	48-009	General Electric	J47, J73, J79 test	1951–1957	
EB-45C	48-010	General Electric	J47 AST	1950	
EB-45C		Allison	J71 testbed	Circa 1954	
B-47B	51-2059/X059	Orenda (Canada)	Iroquois testbed	1956	Canadair CL-52
XB-47D	51-2103, -2046	Boeing/USAF	Turboprop test	Circa 1955	
B-47E	53-2104	General Electric	TF34 development	1969–1974	Display, Pueblo Colorado
B-50A	46-036	Pratt & Whitney	Bomb bay jet test	Tested J57	
B-50A	47-113	General Electric	J73/TPA	June 1954	
NB-52B	52-008	NASA	Shuttle drag chute	1990	B-52 mothership

182   Testbeds, Motherships & Parasites

Aircraft Type	Serial/Registration	User	Purpose	Timeframe	Comments
B-52E	56-0636	Pratt & Whitney	Turbofan testbed		
B-52E	57-0119	General Electric	Jet engine testbed	1966–1972	Hulk at AFFTC
YB-58A	55-0662	General Electric	J93 testbed pod	1959	
RB-66A	52-2828	General Electric	CJ805 testbed	1958–1962	Scrapped at AFFTC
C-8A	63-13686	NASA	AWJSRA STOL	1972–1981	Scrapped in Canada
C-8A	63-13687	NASA	Quick short-haul research aircraft	1978–1990s	NASA #715
C-46A	41-12293	NACA	Icing research	Circa 1943–1947	
XC-69	43-10309	Lockheed	Testbed	1943–1958	Parts salvaged ca 1958
YC-123E	55-4031	USAF	Pantobase landing gear testbed	Circa 1955	
C-124C	52-1069	Pratt & Whitney	T57 testbed	Circa 1956	Nose-mounted propjet
NC-131H	N793VS	Calspan	TIFS	1970–2008	National Museum of USAF
NC-141A	61-2775	USAF	Testbed	1960s–1998	First C-141 built
NC-141A	61-2776	USAF	Testbed	1960s–1990s	Electric Starlifter
NC-141A	61-2777	USAF	Testbed	1960s–1994	"Ash can" tail end
NC-141A	61-2779	USAF	Testbed	1960s–1990s	Used fighter noses
FG-1 Corsair		USN	Ramjet testbed	1944	
F-5E SSBD	BuNo 741519	DARPA/NASA	Testbed	2003	Reduced sonic boom
F-8 DFBW	NASA 802	NASA	Fly-by-wire	1972–1985	
F-8 SCW	NASA 810	NASA	Supercritical wing	1971–1973	Displayed by NASA
F-16 AFTI	75-0750	USAF, Others	Technologies	1978–2000	
F-16XL-1	NASA 849	NASA	Supersonic laminar flow control	1991–1992	
F-16XL-2	NASA 848		Supersonic laminar flow control	1995–1996	
F-16 VISTA	86-0048	USAF	Variable stability	1990s	At Air Force Test Pilot School
F-80A	44-85214	USAF	Ramjets	Circa 1947–1948	Marquardt ramjets
RF-84F	51-1835	General Electric			
XF-84H	51-17059	USAF	Turboprop	1955–1956	Made 8 of 12 flights
XF-84H	51-17060	USAF	Turboprop	1955–1956	Made 4 of 12 flights
XF-88B	46-525	USAF, NACA	Jet plus turboprop	1953–1958	
F-94B	51-5502		Bomarc guidance nose attached		
F-106B	57-2516/NASA 616	NASA	SST nozzle testbed under wings	1966–1971	Mounted two J85 engines
P-38	40-744	AAF	Offset cockpit tests	Circa 1944–1945	Helped validate XP-82
P-51D	44-63528	AAF/Marquardt	Ramjets	Circa 1946	
YP-61	41-18888	P&W	Engine tests	1946–1956	Civil reg N60358
P-61B	42-39498	AAF	Ejection seats	1946	Called Jack-In-The-Box
P-61B	42-39754	NACA	Ramjet testbed	Circa 1947	At NACA Cleveland
P-83			Ramjet testbed	Circa 1946	Aircraft lost; inflight fire
PT-19A	41-20531	Boeing	Subscale B-29 tests		
T-28 Trojan			Yankee extraction test aircraft	1969	
OA-9 Goose	8128	Edo	First hydro-ski test	1948	
JRF-5 Goose		Edo/USN	Second hydro-ski test	Circa 1952	
J4F Widgeon	BuNo 32976	USN/NACA	Seaplane hulls	1947–1948	"Petulant Porpoise"
K-16 (Kaman)	BuNo 04351	USN	Tiltrotor amphibian	1959–1962	Unflown modified JRF
Cessna OE-1		USN/USMC	Hydro-ski testbed	Circa 1953	
SNJ-5C Texan		USN/USMC	Hydro-ski testbed	Circa 1952	
PBM-5 Mariner		USN	Hydro-ski testbed	1955–1958	
PBM-5 Mariner	9158 (?)	USN/Convair	Vertical float test	1963	
U-1A Otter	55-3318		Hydro-ski testbed	Circa 1950s	
X-15A-2	56-6671	USAF/NASA/USN	Modified for ramjet testbed	1964–1967	National Museum of USAF
X-18	57-3078	USAF/Hiller	Tilt-wing testbed	1959–1964	Aircraft scrapped
X-21A	55-0408	USAF/Northrop	Laminar flow control	1963–1964	Used B-66 fuselage
X-21A	55-0410	USAF/Northrop	Laminar flow control	1963–1964	Used B-66 fuselage
Airbus A300	F-BUAD	General Electric	CF6, laminar flow	1984–1992	
Avro CF-100	100760	Pratt & Whitney Canada	JT15D testbed	1960s–1982	
Avro Lancaster	RCAF FM205	Avro Canada	Avro Chinook jet testbed	1951	Scrapped 1956
Avro Lancaster	RCAF FM209	Avro Canada	Orenda testbed	1950–1954	Lost in hangar fire 1956
Avro Lancastrian	RAF VH742		Nene testbed	Circa 1947–1948	Wing nacelle mounted

Aircraft Type	Serial/Registration	User	Purpose	Timeframe	Comments
Avro Lincoln	RA716/G (?)		Theseus propjets	Circa 1947	Also Rolls-Royce Avon jets
Avro Lincoln	RF530 (?)		Naiad propjet in nose	Circa 1948–1956	Also Rolls-Royce Tyne propjet
Avro Vulcan	XA903		Testbed for MRCA	Circa 1971	
BAC-111	N162W	Westinghouse	Avionics testbed		Northrop Grumman later bought it
BAC-111	N164W	Northrop Grumman	Avionics test bed		
BAC-111	N164W	Northrop Grumman	Avionics test bed		
Beech 18	N7927C	Continental	3-engine testbed	Circa 1969	
Beech 18/C-45	RCAF HB109	Pratt & Whitney	PT6 engine in nose	1960–1981	
Beech 18/C-45H	N8640E	Grimes	Lighting testbed	1966–1986	Now Grimes Flying Lab Foundation
Beech 200	BB2/C-GARO	Pratt & Whitney Canada	PT6 testbed	Circa 1979	
Bell L-39	BuNo 90060	USN/NACA	Sweptwing P-63	Circa 1946–1947	Low-speed qualities
Bell L-39	BuNo 90061	USN/NACA	Sweptwing P-63	Circa 1946–1947	Low-speed qualities
Boeing 367-80	N70700	Boeing, NASA	Engines, flaps	Circa 1954–1972	Forerunner of 707
Boeing 707-321	N37681/N707GE	General Electric	CFM56 testbed	1981–1991	Scrapped Mojave 2005
Boeing 720B	C-FETB	Pratt & Whitney Canada	Turbofan, turboprop testbed	1988–2010	National Air Force Museum, Canada
Boeing 720	N720PW	Pratt & Whitney	Turbofan testbed	Retired 2008	
Boeing 720	N720GT/N720H	Allied Signal, Honeywell	5-engine testbed	1980s–2008	Scrapped
Boeing 727-063	N32720	General Electric	GE36 prop fan	1986–1987	
Boeing 737	N35LX	Lockheed-Martin	CATBird testbed	Cooperative Avionics testbed	
Boeing 747-121	N747GE	General Electric	Engine testbed	1992–present	
Boeing 747-446	N747GF	General Electric	Engine testbed	2010–present	
Boeing 747SP	C-FPAW	Pratt & Whitney	Engine testbed		
Boeing 757	N757A	Boeing	F-22 avionics	Circa 1999–present	Wing shape on fuselage
Boeing 757	N757HW	Honeywell	Engine and avionics	2008–	
Bombardier CL-600	N804X	Northrop Grumman	Avionics testbed	2011–	
Convair 990	NASA 710	NASA	Landing research	1993–1994	Retired to Mojave, California
Gloster Meteor	Unknown	Great Britain	Rolls-Royce Soar testbed	Circa 1953	On Meteor wingtips
Gulfstream GII	N650PF	NASA	Propfan testbed	Circa 1987–1988	
Gulfstream GII	N900DH	Safran USA	Silvercrest testbed	2017	RH test engine
Gulfstream V	N99NG	Northrop Grumman	Avionics testbed	2016–	Large ventral "canoe"
Grumman Mallard	Unknown	Northern Consolidated Airlines	PT6 testbed	1964	Northern Consolidated Airlines
Ilyushin Il-14		Motorlet (Czech)	Testbed for chin-mounted propjet	Circa 1970	
JN-4C Canuck		Harlan D. Fowler	Fowler flap test	Circa 1927-28	
Lake LA-4			Hydro-ski testbed	Circa mid-1960s	
Learjet	C-GBRW	Pratt & Whitney Canada	JT15D testbed	Circa 1980s	
Lockheed Vega	N105D	General Electric	Test radar counter-measures	1957–1961	Later restored by David Jameson
Me 110			Jumo 004 testbed	1942	May be first jet testbed
MD-80 (or -81)	N980DC	General Electric	GE36 prop fan	1987–1988	Scrapped 1994
Sikorsky S-55	N419A	Sperry Gyroscope	Instrument tests	Circa 1953	
Sikorsky S-55		Lab for Electronics	Circa 1967	Near Boston, Massachussetts	
Sud Caravelle	N420GE	General Electric	CJ805-23 test	1960–1961	
VC-10	XR809/G-AXLR	Rolls-Royce	RB211 testbed	Circa 1972–1975	Scrapped
Vickers Viscount	CF-TID	Pratt & Whitney Canada	PT6, PW100 test	1982–1988	
Yokosuka P1Y Ginga		Japan	Tsu-11 jet test	1945	

184   Testbeds, Motherships & Parasites

APPENDIX 2
MOTHERSHIP AIRCRAFT

Aircraft Type	Serial/Registration	User	Purpose	Timeframe	Comments
Hot air balloon	Frank Hamilton	John J. Montgomery	Mothership	Circa 1905	Drop glider from altitude
A-26 (JD-1) Invader	Various	USN	Firebee launcher	Circa 1950s	
A-26C (B-26) Invader	43-22494	USAF	XQ-1 mothership	Circa 1950–1951	
A-26 (B-26) Invader	44-35507	USAF	Firebee mothership	Circa 1955	
PB-1W (B-17F)	BuNo 34106	USN	Mothership	Circa 1947	Drop F8F model
B-17G	42-40042	AAF	GB-4 carrier	1944	
B-17G	42-40043	AAF	GB-4 carrier	1944	Operational launch test
B-17G	42-97518	AAF	GB-4 carrier	1944	
B-17G	43-39119 and various	AAF	JB-2 carrier	Circa 1944–1945	Carried two buzz bombs
B-23		AAF	Glide bomb mother	Circa 1942	
B-29	44-62093	USAF	Wingtip tow	1950–1953	Lost in crash
B-29	44-84111	USAF	XF-85 mothership	1948–1949	B-29 named *Monstro*
B-29	45-21748 (?)	USAF	XQ-1 mothership	Circa 1950	Holloman tests
B-29	45-21800	AAF/USAF	X-1 mothership	1946–1948	
B-29/P2B-1S	BuNo 84029, NACA 137	NACA	D558-II mothership		Weeks collection
B-29 (RAF Washington)	Unknown	Great Britain	Model drop		Red Rapier missile
GRB-36D	Several	USAF	Operational FICON	1955–1956	
GRB-36F	49-2707	USAF	FICON test carrier		Also Tom-Tom test aircraft
B-50	Unknown	USAF	Launch XQ-4 drone	Circa 1956	Supersonic drone test
NB-52A	52-003	NASA/USAF	X-15 launch aircraft	1959–	
NB-52B	52-008	NASA/USAF	X-15, varied launcher	1959–2004	Edwards AFB display
B-52H	61-0025	NASA	Intended mothership	2002	Never used
B-52H	60-0036/61-0021	USAF/CIA	D-21B drone	1967–1971	Replaced M-21
PB4Y-2B	Various	USN	Bat glide bomb	Circa 1945	
F7F Tigercat	NACA		Drop aero shapes	1948	
Northrop F-15	45-59300, NACA 111	NACA	Model mothership	1948–1954	Crashed 1968
Northrop P-61C	43-8330	NACA	Model mothership	1951–1954	Smithsonian
Northrop P-61C	43-8336	USN/Martin	Gorgon IV carrier	Circa 1948–1949	
XP-82	44-83887	NACA	Ramjet mothership	Circa 1948–1949	Damaged 1949
F-82B	44-65168, NACA 132	NACA	Ramjet mothership	Circa 1950–1957	USAF's *Betty Jo*
C-5A	69-0014	USAF	Launch Minuteman	1974	In AMC Museum
C-47A	42-23918	USAF	Coupled to Q-14B	1949–1951	Wingtip tow tests
DP-2E Neptune	Various	USN	Firebee launcher		Carried two drones
GC-130A Hercules	57-0496	USAF	Firebee launcher	Circa 1958–2003	DC-130A; also USN later
GC-130A	57-0497	USAF	Firebee launcher	Circa 1959–2007	DC-130A; also USN later
DC-130A	55-0021	USN 158228	Firebee launcher	Circa 1969–1979	Modified for USN 1969
DC-130A	56-0491	USN 158229	Firebee launcher	Circa 1969–1979	Modified for USN 1969
DC-130A	56-0514	USN 560514	Firebee launcher	Circa 1966–1993	Scrapped Mojave 1993
DC-130A	56-0527	USAF	Firebee launcher	Circa 1968–1978	Operational in SEA
DC-130A	57-0461	USAF	Firebee launcher	Circa 1965–2001	USN as 570461
DC-130A	57-0523	USAF	Firebee launcher	Circa 1973–1977	
DC-130E	61-2361		Firebee launcher	Circa 1968–1979	May be AQM-91 carrier
DC-130E	61-2362	USAF	Firebee launcher	Circa 1967–1978	Revert C-130E circa 1979
DC-130E	61-2363	USAF	Firebee launcher	Circa 1968–1979	Revert to C-130E 1979
DC-130E	61-2364	USAF	Firebee launcher	1970–1980	Revert C-130E circa 1981
DC-130E	61-2368	USAF	Firebee launcher	Circa 1966–1978	Revert C-130E circa 1978
DC-130E	61-2369	USAF	Firebee launcher	Circa 1966–1979	Revert C-130E circa 1980
DC-130E	61-2371	USAF	Firebee launcher	Circa 1966–1978	Revert C-130E circa 1979
DC-130H	65-0979	USAF/Lockheed	Firebee launcher		Set lift record in 1976
M-21 (A-12)	60-6940	USAF/CIA	Launch D-21 drone	1964–1966	In Museum of Flight

Aircraft Type	Serial/Registration	User	Purpose	Timeframe	Comments
M-21 (A-12)	60-6941	USAF/CIA	Launch D-21 drone	1964–1966	Lost in use with D-21
747F		USAF	Micro-Fighter carrier	1972–1973	Concept on paper
747	N905NA	NASA	Shuttle carrier	1977–2012	*Enterprise* launcher
747	N911NA	NASA	Shuttle carrier	1991–2012	
747	N744VG	Virgin	Orbit mothership	2015–	Launcher One mothership
Antonov An-225	82060	USSR	Shuttle carrier	1988 first flight	Later carried cargo
Avro Lancaster	RCAF KB848		Ryan Firebee	1955–1957	Nose at Canada Aviation Museum
Avro Lancaster	RCAF KB851		Ryan Firebee	Circa 1957	Struck Off Charge 1961
Curtiss JN-6H Jenny	U.S.		Carry glider	Circa 1918	Launch target glider
Dornier Do 217	Germany		Carry DFS 228	Circ 1945	High altitude recon
Focke-Wulf Fw 56	Germany		Carry DFS 230 glider	Circa 1942	Maintained altitude
Focke-Wulf Fw 190	Various	Germany	Mistel "father" ship		Ferry bomber to target
Heinkel He 111H	Various	Germany	Fi 103 V-1 carrier	1944	
He 274/AAS-1	France		Test mother ship	1945–1953	Scrapped
Klemm KL 35	Germany		Carry DFS 230 glider	1942	Unable to lift glider
Lockheed L-1011	N140SC	Orbital Sciences	Pegasus mothership	1995–	
Messerschmitt Bf 109E	Germany		Carry DFS 230 glider	1943	Combo able to take off
Messerschmitt Bf 109	Germany		Mistel "father" ship		Ferry bomber to target
Porte seaplane	Great Britain		Launch Bristol Scout	1916	Range extender
Stratolaunch	Vulcan Aerospace		Largest mothership	2017	Under development
White Knight One	N318SL		Scaled Composites		*Space Ship One*
White Knight Two	N348MS		Scaled Composites		Sometimes called Eve
Short Model 21 Maia	Great Britain		Launch mailplane	1938	Airmail range extender
USS *Akron*	Model ZRS-4	USN	Carry/launch aircraft	1932–1933	Lost at sea
USS *Los Angeles*	Model LZ.126/ZR-3	USN	Carry/launch aircraft	1929	Tested concept
USS *Macon*	Model ZRS-5	USN	Carry/launch aircraft	1933–1935	Lost off California coast
Zveno Project		USSR	Carry/launch aircraft	1931–1942	Various bomber motherships carrying fighters
SPB		USSR	Operational Zveno TB-3 mothership	1937–1942	Launched I-16s as dive bombers

APPENDIX 3
GLOSSARY OF ACRONYMS AND ABBREVIATIONS

AAF: Army Air Forces (of the United States)
AEDC: Arnold Engineering Development Center
AFB: Air Force Base
AFFTC: Air Force Flight Test Center (at Edwards AFB, California)
AFTC: Renamed Air Force Test Center, formerly AFFTC (see above)
AMC: Air Mobility Command (of USAF)
BuAer: U.S. Navy Bureau of Aeronautics; U.S. Navy aircraft serial numbers are sometimes referred to as "BuAer numbers" (See also BuNo)
BuNo: U.S. Navy aircraft serial numbers are sometimes referred to as "BuNo"
CFD: Computational Fluid Dynamics
CRADA: Cooperative Research and Development Agreement
DFRC: Dryden Flight Research Center (NASA)
DARPA: Defense Advanced Research Projects Agency
FICON: FIghter CONveyor
FRR: Flight Readiness Review
GE: General Electric
GTOW: Gross Take-Off Weight

IAS: Indicated Airspeed
LFC: Laminar Flow Control
NACA: National Advisory Committee for Aeronautics
NAS: Naval Air Station
NASA: National Aeronautics and Space Administration
NMUSAF: National Museum of the United States Air Force
P&W: Pratt & Whitney
RPV: Remotely Piloted Vehicle
RR: Rolls Royce
SSBD: Shaped Sonic Boom Demonstration
STOL: Short Takeoff and Landing
TAS: True Airspeed
TIFS: Total In-Flight Simulator
UAV: Unmanned Aerial Vehicle
USAF: United States Air Force
USMC: United States Marine Corps
USN: United States Navy
VISTA: Variable Stability Inflight Simulator Training Aircraft
VTOL: Vertical Takeoff and Landing

APPENDIX 4
X-15A-2 REPORT

A 1969 official report written by Johnny G. Armstrong, a longtime Air Force Flight Test Center (AFFTC) engineer who was key to the X-15 program, captures the decisions and drama of the stretched X-15A-2 testbed aircraft program. Armstrong's report is noteworthy for its professional insight and clarity. That report, AFFTC Technology Document No. 69-4, is excerpted here, unedited.

Foreword

This Technology Document presents a general history of the X-15-2 airplane and a detailed summary of the flight planning and conduct of the test program on the modified X-15A-2. The test program of the modified aircraft was conducted at Edwards AFB, California, from 25 June 1964 to 3 October 1967 through the joint efforts of the NASA Flight Research Center and the Air Force Flight Test Center. The aircraft was flown by AFFTC pilots, Colonel Robert A. Rushworth; Major William J. Knight; and Mr. John B. McKay of the NASA Flight Research Center. The actual envelope expansion program consisted of eight flights between November 1965 and October 1967. When this report was published, the aircraft was being transferred to the Air Force Museum at Wright-Patterson AFB, Ohio.

Abstract

After-having been extensively damaged during an emergency landing on its thirty-first flight (November 9, 1962), the X-15-2 aircraft was rebuilt and modifications were incorporated to increase the vehicle performance capability to allow flight testing of a hypersonic ramjet engine. The increased performance was derived from additional propellants contained in two external drop tanks. A total of 22 flights were made with the modified aircraft which had been redesignated the X-15A-2. The initial flights were to evaluate the handling quality changes resulting from the modification. The modified propellant system with the external tanks was satisfactorily developed on a ground test stand and performed adequately during flight. Although successful ejection of the external tanks occurred on their separation from the aircraft on each flight, carrying the tanks imposed new constraints on flight planning such as tank ejection flight limits, tank impact locations, and revised emergency lake requirements.

The ablative material developed to protect the aircraft against temperatures exceeding the original aircraft design appeared to perform satisfactorily on the two fully coated flights flown. On the last flight of this aircraft, the vehicle achieved a maximum Mach number of 6.7 (without using all the propellants available). Extensive heat damage was encountered on the dummy ramjet and lower ventral fin as a result of unexpected increased heating rates due to shock impingement and flow interference effects. While the aircraft was being repaired, the X-15A-2 program was terminated and the maximum speed capability of the aircraft was never achieved.

The type of problems encountered during the course of the envelope expansion program may well be expected on other vehicles operating in the speed regime where aerodynamic heating will be an appreciable factor. For instance, the test program was slowed by premature landing gear extensions during flight as a result of aerodynamic heating. These failures should serve as a warning of the potential problems that could occur as a result of minor modifications when operating in a high temperature environment.

In addition to the continued demonstration of piloted landing of an unpowered low L/D vehicle, other techniques developed during the program are applicable to orbital lifting re-entry vehicles: application of and flight with an ablative coating, protection of a canopy window with a pilot-actuated covering, and development of an extendable pitot tube as an airspeed source for the terminal landing maneuver.

Introduction

The X-15A-2 obtained a maximum Mach number of 6.7 on October 3, 1967. At the time this flight occurred the X-15A-2 was involved in an envelope expansion program to extend the maximum Mach number capability from Mach 6 to approximately Mach 8. The aircraft was then to be used as a flying testbed for testing a hypersonic ramjet engine. Financial cutbacks following this flight resulted in termination of X-15A-2 from the active flight program without ever having realized the aircraft's maximum velocity capability.

General Aircraft History

The X-15 free flight program began with a glide flight of X-15-1 on June 8, 1959. X-15-2 entered the flight program on September 17, 1959, making the second X-15 flight and the first powered flight with the XLR11 engine. Two XLR11 engines were used to power each of these two aircraft to begin the envelope expansion program of the X-15 before the XLR99 engine became available. . . The X-15-2 made a total of nine flights with this engine configuration and flew the first flight with the XLR99 engine on November 15, 1960. During 1961 the aircraft was involved in the envelope expansion program with the XLR99 engine, making a total of nine flights that year and obtaining a maximum altitude of 217,000 feet and a maximum Mach number of 6.04.

Ten flights were made with the X-15-2 in 1962 with the majority of the flights being designed to obtain aerodynamic heating data at the high Mach number and high dynamic pressure under quasi-steady conditions. During the latter part of 1962, flight tests were conducted to verify a predicted improvement in lateral directional handling qualities at high Mach numbers and high angles of attack with the lower movable ventral fin removed. The thirty-first flight of X-15-2 on November 9, 1962, was planned to further investigate the aircraft's ventral-off handling qualities. However, during this flight only 30-percent thrust could be obtained from the XLR99 engine as a result of a throttle control failure. An emergency landing was attempted at the launch lake (Mud Lake) in accordance with preplanned alternate procedures. At touchdown the left main gear strut collapsed, causing the aircraft to skid sideways and turn over on its back. The gear structural limit was exceeded primarily because landing flaps failed to extend. The aircraft suffered extensive damage. The aircraft had accumulated a total free flight time of 4 hours, 40 minutes, 32.2 seconds at this point. A decision was made to rebuild the aircraft in order to complete the planned experiments and to incorporate modifications to increase the performance of the aircraft and thus allow it to be used as a testbed for a hypersonic ramjet engine. Approval was given under Contract AF33(657)-11614 for North American Aviation, on May 13, 1963, to proceed with repair and modification of the aircraft at a cost of approximately 5 million dollars. The modified aircraft . . . was returned to Edwards AFB on February 19, 1964, and made its first flight June 25, 1964.

Description of Aircraft

Numerous changes were made to the X-15-2 while it was being rebuilt. The major change was the addition of two external propellant tanks. These jettisonable external propellant tanks were designed to increase the engine burn time by approximately 70 percent, thereby increasing the performance capability of the aircraft required for testing the ramjet engine.

The external tanks were each approximately 23.5 feet long and 38 inches in diameter.

The left-hand tank . . . weighing 1,150 pounds empty, contained three helium bottles required for propellant tank pressurization in addition to a capacity for approximately 793 gallons of LOX. The right-hand tank . . . weighed 648 pounds empty and contained approximately 1,080 gallons of anhydrous ammonia. The total weight of additional propellant to be carried in the external tanks was approximately 13,500 pounds. Because of the difference in empty weight and in propellant volumes, the left-hand tank was approximately 2,000 pounds heavier than the right at launch.

The external tank jettison system . . . contained two sets of fore and aft gas cartridges to eject the tanks from the aircraft. In addition, the design included a solid propellant sustainer rocket on the nose of each tank to impart a nose-down moment upon jettison to improve separation characteristics at supersonic speeds. For a normal empty tank jettison both sets of gas cartridges were fired and the nose rocket was ignited. In the case of a requirement to make an emergency tank ejection while the tanks were still full, only one set of the gas cartridge ejectors were fired and the nose rocket was not activated.

The high cost of these tanks dictated that they be reusable, hence each tank contained its own recovery system consisting of a drogue and descent chute. The drogue chute was deployed immediately after separation and the main descent chute deployment was initiated by a barometric sensor normally set for 8,000 feet.

A 29-inch extension was added to the fuselage in the area of the center of gravity between the LOX tank and the anhydrous ammonia tank.

Tanks containing 48 pounds of liquid hydrogen for the ramjet engine were to have been installed in this area.

Additional hydrogen peroxide required for the extended engine propellant pump operation was stored in tanks in the extended aft side fairings. A helium tank for additional propellant pressurization gas was installed on the aft fuselage above the engine.

The design included a longer landing gear that would provide ground clearance for landing with a ramjet engine installed. Since the ramjet engine was not to be available until much later in the test program it was decided to take advantage of the increased landing load margin that could result from a shorter main gear during the initial portion of the test program. The strut of this interim gear was 6.75 inches longer than the standard X-15 gear.

Drawing from the experience of the initial X-15 envelope expansion program when the standard windshield design suffered several glass fractures caused by thermal stress near the corners of the rectangular glass retainer, the X-15A-2 windshield was designed with an elliptical shape. In addition, three panes of glass were installed in the new design instead of 2 panes as in the normal X-15.

To protect the aircraft structure from the high aerodynamic heating in the Mach 6 to 8 regime, an ablative material was chosen to cover the aircraft.

To facilitate projected experiments, two additional modifications were made to the aircraft. A "sky hatch" was added just behind the cockpit, which featured doors that could be opened upon command from the pilot near peak altitude on high-altitude flights. The ultraviolet stellar photography experiment (Star Tracker) was later installed in this compartment. The right-hand wing tip was designed to be removable. This removable wing tip was to have allowed testing of advanced materials and/or structural design.

The onboard instrumentation recorders utilized during the envelope expansion program consisted of five 36-channel oscillographs, eight 3-channel oscillographs, two 14-track tape recorders, one 24-cell manometer recorder and one cockpit camera. In addition, an 86-channel PDM telemetry system was used to transmit parameters in real time from the aircraft.

Test And Evaluation

Summary of Initial Flights After Modification

Wind-tunnel tests of X-15A-2 were conducted in the summer and fall of 1963. The tests indicated that very little difference existed in aerodynamic characteristics between the modified X-15 without external tanks and the standard X-15. The movement of the normal flight cg 10 percent forward apparently compensated for the destabilizing effect of extending the fuselage 29 inches forward and thus resulted in little change to the static stability derivatives. The low level of directional stability of the standard X-15 at Mach 3 and 12 degrees [alpha] was slightly lower for the modified aircraft. However, at re-entry conditions of Mach 5 and 20 degrees [alpha] the dihedral effect still remained favorable.

The longitudinal trim characteristics of the modified aircraft remained the same as for the standard X-15 for Mach numbers less than 4 and stabilizer deflections less than 15 degrees. At higher Mach numbers . . . the modified aircraft's trim capability was about 5 degrees less in angle of attack.

The initial flights of X-15A-2 were planned to obtain stability and control maneuvers to verify the wind-tunnel predictions. Data obtained on the flights did in fact verify the wind-tunnel results; however, the verification program took longer than expected when trouble was encountered with the modified landing gear system. On the second flight (2-33-56) of the modified aircraft, after obtaining a maximum Mach number of 5.23, the nose gear inadvertently extended at Mach 4.4. Despite the degraded control and increased drag resulting from the extended gear, the pilot was able to return to Rogers Dry Lake at Edwards Air Force Base. The chase aircraft pilot was able to verify that the nose gear appeared to be structurally sound and in the locked position but that the tires showed heat damage. The pilot elected to attempt a landing, which was accomplished normally except that both nose gear tires blew out on landing. Investigation revealed that aerodynamic heating was the cause of the failure; specifically, the expansion of the fuselage was greater than the capacity of the tension regulator/temperature compensator device of the gear release cable. This caused an effective pull on the release cable, which then applied a load on the uplock hook. An additional load on the uplock hook was imposed by an outward bowing of the nose gear door. The load from both of these sources caused the uplock hook to bend, allowing the gear to extend. This failure was duplicated by ground tests simulating the fuselage expansion and by applying heat to the nose gear door. The key linkages were redesigned and the system was subjected to the same ground heat test without failure. The same mission plan for stability and control data was planned for the next flight (2-34-57). Again, shortly after shutting down the engine at a maximum Mach number of 5.2, the pilot experienced a similar noise and aircraft trim change at Mach 4.5. The small nose gear scoop door had extended. This door in the normal gear extension sequence was used to impose airloads on the nose gear door to assist in the extension of the nose gear. Although not as serious a failure as that of the previous flight, it again precluded obtaining dampers off stability data.

The nose gear door was redesigned to provide positive retention of the scoop door regardless of the thermal stresses.

A slower speed flight (2-35-60) was flown next to a maximum Mach number of 4.66 to check out the modifications on the nose gear door. The nose gear performed normally and additional stability data were obtained.

On the next flight (2-36-63) the right main gear extended at Mach 4.4. Again the chase pilot was able to verify that the gear appeared structurally sound and a normal landing was made. Postflight inspection revealed that the uplock hook had bent allowing the gear to deploy. Again, the source of the high load on the uplock hook was concluded to be from aerodynamic heating. [T]he temperature gradient between the inside and outside of the stowed gear and strut resulted in differential expansion causing the gear to bow in the middle. The additional length of the interim gear caused its thermal load to be almost twice as high as that of the standard gear. Thus when the critical temperature was reached in flight, the load became sufficiently large to deform the hook causing the gear to extend. The uplock hook was redesigned and pertinent instrumentation was added to allow flight evaluation of the change. To safely test the modification in flight, a "temperature profile" was selected with heating rates similar to the flight when the failure occurred but a total temperature less than that at which the failure occurred. The normal procedure used to predict the temperature on the vehicle was to fly a profile on the X-15 analog simulator and then load the pertinent data into a digital program that calculated temperature at selected points on the vehicle. To obtain a discrete temperature versus time profile would be an iteration process that could take weeks to accomplish. However, the use of a real time temperature simulation of the X-15A-2 . . . made the task relatively simple. A plot of the desired temperature time history was placed on an X-Y plotter and the flight planner was able to observe the temperature as he simulated the "flight" to assess the match and make immediate adjustments to the flight path until the desired profile was obtained. The practicality of the resulting mission was also evaluated by simulating off-design flights to determine which flight conditions could result in undesirable temperature overshoots.

Five more flights (2-38-66 through 2-42-74) were flown before the envelope expansion program was begun. These flights continued the study of stability and control and landing gear performance tests. Three of the flights were primarily to obtain data for the ultraviolet Star Tracker experiment. However, little usable star tracking data were obtained because of problems incurred in maintaining the precise attitudes required for the experiment.

During successive flights, attempts were made to improve the reaction augmentation system (RAS) which provided rate damping about all three axes with the reaction control system to assist the pilot in maintaining the required aircraft attitudes.

The Star Tracker flights were discontinued because of the position of the desired stars at that time of the year.

Preparation for Envelope Expansion Flights

Prior to the arrival of the modified aircraft and concurrent with the initial flight phase, studies were made of the unique problems associated with flying the aircraft with external tanks.

External Tank Impact Area

The addition of external tanks to the X-15 added an additional constraint to the flight planning task, that of having the aircraft over a satisfactory tank impact area at the time of planned tank ejection. To define the area of probable tank impact, trajectory calculations were made using the following conditions as the standard for tank ejection.

It was assumed in these calculations that the tanks would fly at zero angle of attack, i.e., zero lift. Drag coefficients were estimated for the tanks and for the 33.2 sq ft drogue chute. A wind drift allowance was included while descending with the main chute deployed in the direction of the predominate winds for this area.

This study defined a ground recovery area of 8.1 NM by 9 NM. The legal clearance to drop the tanks in the geographic area defined by this study had to be obtained.

Additional areas were established for a failure of the drogue chute and for an emergency tank ejection immediately after X-15 launch. Although it was not feasible to obtain land rights to drop tanks in the entire area so defined because of the large land area involved, the possibility of an impact in these areas was definitely considered in selecting the ground track of the X-15 with external tanks installed. The approach taken was similar to that of operational aircraft flying with external stores, namely, that a drop at an unplanned location would be the result of some malfunction or emergency.

Emergency Lake Coverage

The X-15 flights were planned with the requirement that the aircraft always be within gliding distance of a dry lake suitable for landing. Thus the X-15 was launched within gliding distance of a "launch lake" and during its flight back to Edwards passed by several dry lake beds that had been tested and marked with runways for X-15 landing. During the entire X-15 program, 10 landings were made at these remote lake beds.

Since the initial acceleration of the X-15A-2 with external tanks was considerably less than that of the standard X-15, it was necessary to re-evaluate the emergency lake coverage for flights with external tanks. A parametric simulator study was made to determine the glide capability of the aircraft for different engine burn-time along the design profile to 100,000 feet.... Placing the geometry of the existing emergency lakes at their respective positions on the distance scale makes possible a quick analysis of the emergency lake coverage available. This analysis concluded that, of the existing launch points, only a launch from Mud Lake was suitable for flight with external tanks. However, since the Mud Lake launch point is only 187 nautical miles from Edwards, a more distant launch point would have been required for flights with maximum velocity approaching 7,000 fps. The use of Smith Ranch as a launch point was desired but unfortunately the distance between Smith Ranch and Mud Lake was too great for the glide capability of the aircraft and a time period existed when the aircraft would have been without a suitable landing site. It was hoped that a usable lake could be found between Smith Ranch and Mud Lake to fill the time gap that existed. An uprange survey of dry lakes yielded no such usable landing site. However, a relatively large dry lake, Edwards Creek Valley, approximately 15 NM northwest of Smith Ranch was found to be suitable. The emergency lake coverage from an Edwards Creek Valley launch, although not as good as desired (at least a 20,000-foot high key), did provide for a straight-in approach to Smith Ranch and Mud Lake if an engine shutdown occurred at the most critical time. In addition, for an emergency occurring at the time of tank ejection, a landing could have been performed at Smith Ranch. Upon completion of the study that proved Edwards Creek Valley to be suitable and required for use as part of the envelope expansion flight program, considerable coordination was required to obtain the right to use the lake as an emergency landing site and to obtain approval to drop the tanks in a specified area near Smith Ranch.

External Tank Separation

The utilization of external tanks on the X-15 was unique in that the tanks had to be ejected from the aircraft. Structural limitations of the aluminum tanks and degrading handling qualities dictated that the maximum allowable Mach number with the external tanks be 2.6. Prior to reaching that speed the tanks had to have separated from the aircraft cleanly. A recontact with the aircraft could have possibly resulted in immediate catastrophic failure or apparent minor local damage that could later become catastrophic as high temperatures were encountered. A normal landing with the tanks installed was not possible because of the increased drag and lack of ground clearance. Hence, considerable effort was expended to assure adequate separation characteristics of the tanks from the aircraft.

Theoretical analyses were made of the separation characteristics based on force and moment wind-tunnel data obtained with the tanks in the vicinity of the X-15 model. Dynamic tank ejections were also made in the wind tunnel. Good agreement between the two methods of analysis were obtained. The velocity and pitch rate imparted to the tanks by the ejector system were determined from qualification tests. Based on these tests and analysis, the ejection boundary ... was established; between +10 and -2 degrees angle of attack and at dynamic pressures less than 400 psf. However, simulator studies showed that the major portion of the planned profile from launch to depletion of the external propellants ... was outside the allowable ejection boundary and that precise control would be required to achieve the satisfactory ejection conditions. A re-analysis of the data indicated that acceptable, although not as good, separation characteristics would probably exist at dynamic pressures up to 600 psf and that this increased boundary could be verified from results of the initial tank ejections.

Prior to the first tank flight, two dummy tank ejection tests were performed from the aircraft. The aircraft was placed over a 10-foot deep pit. Beams with similar mass and inertia properties were constructed to simulate the empty tanks. Preloaded cables attached to the beams applied simulated aerodynamic drag and side loads. A single set of ejector cartridges was used on the first test at simulated air loads of 400 psf dynamic pressure, 5 degrees angle of attack and 3 degrees sideslip. The second test used both sets of ejector cartridges at a simulated dynamic pressure of 600 psf. Both tests were successful and high-speed motion pictures showed good separation characteristics. During the tests the hydraulic and electrical power was supplied by the aircraft and the SAS system was engaged to assure that no detrimental effects on the aircraft system occurred during the simulated tank ejection.

During a Design and Operating Criteria Review of the modified aircraft, concern was expressed for the separation characteristics of the external tanks with the tanks partially full. The pilot could have found himself in such an emergency situation requiring the ejection of the tanks with a partial load of propellant if the engine shut down prematurely within the first 60 seconds of flight. The tanks and the ejection system were designed for only a full or empty tank ejection. It was considered that the tanks with the ejection system as initially designed would not withstand the loads imposed during ejection with partial fuel. Studies were initiated to find a suitable solution to this possible problem area. Three separate approaches were studied as follows:

1. A rapid external propellant dump system that would empty the external tanks in 15 seconds.
2. A system of tank baffles that would reduce fuel slosh.
3. A rapid fill system that would allow the external tanks to be filled from the internal tanks.

Each of these schemes had its own advantages and disadvantages, but all complicated the system design further and required excessive time before the completed hardware could be designed, constructed, and qualified for flight. After much study and consideration it was decided that the flight program should be begun with the tanks as initially designed. This calculated risk was in part considered reasonable because to that date the XLR99 engine had not encountered a premature shutdown from 100-percent thrust after a successful light was obtained after launch. However, the study did bring forth one design change to the tank ejection system that was incorporated. A third ejection button was added to the cockpit for ejection of both external tanks with a partial load of propellant. In order to reduce the loads imposed on the tanks at ejection, the new button activated only one set of the ejection cartridges and also caused the separation nose rocket to fire.

Handling Qualities

The X-15 simulator was updated with the wind-tunnel determined derivatives of the X-15A-2 with the lower ventral installed and with external tanks installed. A complete assessment of the predicted handling qualities of the aircraft in this configuration was performed on the simulator. A lower level of longitudinal static stability existed with the external tanks installed making the aircraft considerably more sensitive to control.

The overall control task was further complicated by the offset center of gravity caused by the external tanks . . . At launch, the vertical cg was approximately 9 inches below the aircraft center line, and became less as the external propellants were consumed. This offset below the thrust vector resulted in a nose-down pitch at engine light that had to be counteracted with additional nose-up stabilizer trim. The left lateral displacement of the cg caused by the heavier LOX tank and propellants, resulted in a left rolling moment which had to be counteracted by the pilot with right aileron.

As was the case with the basic aircraft, the poor handling qualities at the high angles of attack was due primarily to the large negative dihedral effect . . . caused by the presence of the lower ventral fin. For a yaw damper failure with the speed brakes out . . . a divergent sideslip oscillation persisted above about 6 degrees angle of attack. Although the divergence could be damped by the pilot with rudder inputs, continuous attention to the task was required. The simulator also showed that the divergent yaw oscillation could be eliminated by turning off the roll damper, however, the pilot would then have to fly the aircraft with less lateral-directional stability. From the simulator studies it was determined that, because of the relatively low altitude profiles required, the aircraft could be safely flown after a roll and/or yaw damper failure by maintaining an angle of attack of less than 8 degrees. For the initial envelope expansion flights, this characteristic could be accepted and attempts would be made to obtain flight verification. For the projected ramjet tests, where flight at high dynamic pressure would be required, a divergence of this type could have been too rapid for the pilot to take corrective action. Hence, it was deemed desirable to provide a redundant yaw damper, and design of an alternate yaw damper similar to that existing in the pitch and roll axis was begun.

Ablatives

The initial design of the modified aircraft included an ablative material to protect the aircraft structure. This material was later considered unacceptable for use on the X-15. The principal objections to the original material were a cure-cycle requiring a heat of 300 degrees F, a water solubility problem, and poor thermal protection efficiency.

In late 1963 a joint NASA-USAF committee was formed to select an ablative material suitable for the X-15A-2 application. Initially a large number of materials were screened with respect to the following areas: shielding effectiveness, room-temperature cure-cycle, bond integrity, operational compatibility and refurbishment. The total weight of the thermal protection system was not to exceed 400 pounds. The initial testing of candidate mater[i]als was accomplished in the 2-inch arc jet at the University of Dayton Research Institute. Flight tests of sample materials at different locations on the X-15's were valuable in uncovering differences in the materials under actual flight environment; particularly in terms of application techniques, bonding effectiveness, and resistance to aerodynamic shear loads. Final evaluation was accomplished at the NASA Langley Research Center's 2500 KW arc jet under heating conditions simulating peak heating rates expected on the X-15A-2 maximum velocity mission. Four ablative materials qualified for the X-15A-2 application. A request for proposal was sent to the manufacturers of the materials. In late 1965 a contract was let to the Martin-Marietta Company to design and apply a sprayable silicone ablator to the aircraft. The basic ablative material was designated MA-25S and had a virgin material density of 28 pounds per cubic foot. The material could be sprayed and cured on the aircraft at room temperature (70 to 100 degrees F). Special premolded fiber reinforced silicone material (ESA3560-IIA) . . . was designed for all leading edges. A premolded flexible material (MA-25S-1) was developed to cover seams of access panels required for preflight activities. This ablative material as well as all the other candidate materials was known to be impact sensitive in the presence of LOX. Tests showed that a local detonation would occur on the material submerged in LOX when struck with a force of 8.5 foot-pounds. Special precautions were taken to prevent contamination of the aircraft systems by the ablative material. The interior of the aircraft was protected during application by masking off all openings into the aircraft and contamination measuring devices were installed in the interior to verify the protection. Filters were installed in the propellant line for inflight protection. An ablative sealer (DC90-090) was applied over the final coat to prevent flaking off of the ablative material during maintenance and pre-flight preparation. In addition, this rubbery white sealer decreased the LOX sensitivity to 26 foot-pounds.

Canopy Eyelid

During the arc tests it was observed that loosened material from the ablative surface tended to reattach to surfaces downstream of the test specimen. Flight tests were performed with a panel of X-15 windshield glass mounted on the vertical tail aft of a sample patch of the ablator. The glass panel opaqued, which could have restricted the pilot's vision. Three different concepts were considered to protect the canopy windshield:

1. Explosive fragmentation of the outer windshield glass.
2. Boundary layer blowing over the windshield area.
3. Hinged metal shield (eyelid).

The eyelid was chosen as the most practical method and the design was incorporated onto the aircraft's left windshield. The eyelid was to be closed prior to launch and not to be opened again until the aircraft approached the landing site at speeds less than Mach 3.

Pitot-Static System

The standard pitot-static pickups had to be relocated and redesigned because of the presence of the ablative material. The standard static source was located on the side of the forward fuselage which would be surrounded by ablative material. A vented compartment behind the canopy was chosen as a static location and found to be suitable during flight tests. The standard dog-leg pitot tube ahead of the canopy was to be replaced by an extendable pitot . . . since temperatures above design would have been experienced at high Mach numbers. This tube remained within the fuselage until the aircraft decelerated below Mach 2. The pilot then actuated the release mechanism and the tube extended into the airstream.

Aircraft Configuration Changes

After the successful demonstration of the advanced design on Flight 2-50-89, the envelope expansion program was reoriented. In order to preclude the requirement for separate envelope expansion programs for different aircraft configurations, it was decided to reconfigure the aircraft to its final aerodynamic configuration.

During the following winter rainy season when the X-15 could not fly because of water on the lake beds, preparations were made for this new phase of the program. A "dummy" ramjet shape was designed to be mounted in place of the lower movable ventral. A parachute recovery system was included in the design to allow the unit to be recovered after jettison on the landing approach. Since some refurbishment would be required after flight, three of the dummy ramjets were constructed. Forty-two inches of the forward part of the fixed ventral were cut off and the remaining portion of the ventral configured as a ramjet pylon. Other configuration changes were the installation of the canopy eyelid, installation of Yaw ASAS, and the removal of the ballistic control system rockets.

Limited wind tunnel data were obtained on the basic ramjet configuration without external tanks at Mach numbers of 1.5, 3.0, and. 6.5.

Incremental effects of replacing the existing movable ventral with the ramjet were determined from the wind tunnel tests and this increment was applied to the ventral off data in the simulator mechanization. The increment at 1.5 Mach number was assumed to apply at all lower Mach numbers in the simulation. Wind tunnel data with the tanks on were obtained at only 1.5 Mach number. The stability derivatives for this configuration were mechanized by adding the incremental differences to the existing ventral on derivatives based on the increment at 1.5 Mach number. The lack of wind tunnel data below Mach 1.5 resulted in some degree of uncertainty in simulator validity at transonic and subsonic speeds. However the handling qualities determined on the simulator were satisfactory even when the derivatives were degraded by 30 percent. The predicted lateral-directional handling qualities were essentially the same as with the ventral on. However, the longitudinal trim characteristics were predicted to be quite different, particularly at low angles of attack. The additional frontal area of the ramjet located below the cg resulted in a nose down pitching moment that required additional nose up stabilizer. This predicted longitudinal trim difference was particularly apparent on the simulator in that frequent nose up trimming was required to maintain zero normal acceleration as the Mach number increased.

Flight No. 2-51-92

Evaluation of the aircraft's handling qualities with the dummy ramjet engine installed was the main purpose of this flight flown on 5 May 1967. In planning the flight it became necessary to consider a new constraint. The canopy eyelid was designed to be used when the aircraft was coated with ablative material and thus to keep the temperatures below the design level, the eyelid itself had to be covered with ablative material. This presented the problem that undesirable thermal stresses could result from temperature differentials between the cooler structure behind the ablative coated eyelid and the unprotected structure on the canopy below the eyelid. The real-time temperature simulation was used to establish the flight plan which best satisfied the objectives of the flight within the constraint imposed by the canopy temperature limitation. The resulting flight plan limited the maximum velocity to 4,500 fps and thus dictated a launch closer to Edwards than Mud Lake. The flight was flown from the Hidden Hills launch point, which is 121 NM from Edwards.

During the approach to high key, the pilot opened the canopy eyelid at approximately 2.2 Mach number. This opening was accompanied by a slight trim change in all three axis (nose up, roll right, right yaw). An abrupt nose up longitudinal trim change occurred when the aircraft was decelerating in the transonic range. This was unexpected since no wind tunnel data had been obtained below Mach 1.5; however, the pilot was able to trim nose down at a sufficient rate to counteract the trim change.

On final approach the pilot jettisoned the ramjet, and there were no aircraft trim changes associated with the event. Ramjet separation characteristics were calculated by the contractor and under certain conditions the ramjet could recontact the aircraft, therefore a recommended ejection envelope and cg limits for the unit were established. To obtain data on the actual ramjet separation, a mobile tracker with telescopic cameras was located normal to the final approach track at the point of expected ramjet ejection. The pictures obtained were excellent and detailed analysis of the film was possible. The agreement of the actual separation with the theoretical calculations was good, with the actual clearance being better than predicted. The ramjet recovery system operation was unsatisfactory. The stranded cable attaching the recovery parachute to the ramjet failed as the chute deployed: however the impact damage was not major and the unit was refurbishable.

Flight No. 2-52-96

After Flight 2-51-92 on May 8, 1967, preparations were begun to configure the aircraft for the first flight with a complete ablative coating.

The application of the ablative coating was begun on 25 May and was completed in five weeks requiring approximately 2,000 man-hours. The design of the premolded gloves covering the leading edge of the wings, tail surfaces, and canopy was based on a Mach 8 design mission. The application of the sprayed on ablative material was based on the expected heating rates on a more realistic maximum design mission to 7.4 Mach number and the established requirement to limit the undersurface temperature to 600 degrees F because of loss of bond strength at higher temperatures. The thickness varied from 0.65 inches at the leading edge of the horizontal stabilizer to 0.02 inches at locations on the upper surface of the wings.

One of the uncertainties of the flight plan was the amount of performance degradation caused by the ablative material. The increase in the drag coefficient caused by the increased leading edge radius, trailing edge thickness and skin friction was theoretically estimated to be 0.015. This increment was introduced into the simulator as part of the data on which the flight plan was based. During pilot preparation on the simulator, this parameter was varied to acquaint the pilot with possible deviations from the planned profile. To acquaint himself with possible changes in energy management at landing pattern speeds due to a reduction in L/D, the pilot practiced the X-15 approach in the F-104 at less than normal X-15 L/D.

In considering the results of possible system failures, it was recognized that a failure of the pilot's attitude indication at high speed could possibly leave the pilot without adequate roll reference. Outside reference would not be available because the left window would be covered with the eyelid, and the aircraft flight conditions would be outside the limits for opening the eyelid. Also, the right window could become coated with ablative residue. Therefore an F-104 attitude system was installed which included a two-inch indicator in the cockpit panel.

Prior to the flight, a planned captive flight was made to check the aircraft systems after coldsoak at altitude. This check was deemed desirable to determine if the presence of the ablative coating on the aircraft changed the environment inside the aircraft enough to affect the aircraft systems. The external tanks were installed on the aircraft for this captive flight. Although some aircraft discrepancies were discovered and later corrected they were not attributed to the environment. The external tanks were removed prior to flight 2-52-96.

Flight 2-52-96 was flown on 21 August 1967. A maximum speed of 4,939 fps was obtained at engine burnout. The boost profile from Hidden Hills was flown essentially as planned. After burnout the pilot performed a series of stability and control maneuvers. No differences in handling qualities were detected that could be attributed to the presence of the ablative material.

While approaching the pattern, the pilot actuated the alternate pitot tube. After the aircraft had slowed to a subsonic speed, the pilot compared the indicated airspeeds from the alternate pitot-static sources with the standard system and airspeeds called from the chase aircraft. The airspeeds from the alternate sources were 50 to 70 knots higher than those from the standard sources, but showed the closest agreement with the chase aircraft; therefore, the airspeed from the alternate system was used for landing. On the previous flight without the ablative coating, the difference between the alternate and standard airspeeds had been noted to be only 20 to 30 knots.

The performance of the ablative material on this relatively low speed flight was very good. Most of the degradation occurred on the leading edges while remaining areas showed little change in appearance. Two localized problems occurred during the flight. Small 1.5-inch sections of the ablative material separated from the left side of the upper vertical stabilizer. This separation occurred at the interface of two separate spray coatings. The failure was attributed to the spray mixture being too dry at the start of the second coat. A tensile test of the ablative coat in this area prior to the flight had indicated that the adherence was slightly sub-marginal, but was not considered critical for this flight. The second problem was erosion of the ablative material on the leading edge of the ramjet pylon. . . . *It was believed that this increased erosion was the result of shock wave impingement from the ramjet; however, the seriousness of the problem was not fully appreciated until the next flight.* (Emphasis added.)

Flight No. 2-53-97

With the successful demonstration of satisfactory handling qualities with the ablative coating on the aircraft and effectiveness of the ablative material as a heat shield, the aircraft was cleared for the Mach 6.5 envelope expansion flight.

Aircraft preparation for this flight consisted mainly of refurbishing the ablative material wear resulting from the last flight. This task required approximately 700 man-hours over a two-week period.

Key piloting events [were] associated with trajectory control to achieve desired conditions for tank ejection and then to attain 100,000 feet. The planned maneuvers after shutdown were to verify the stability and control derivatives and to establish longitudinal trim characteristics. Extension of the speed brakes after shutdown was for energy management purposes. Note that an allowance for the increased drag due to the ablative material was included in the simulation. Initially a drag coefficient of 0.015 was applied; once the velocity increased to 5,500 fps, it was assumed that the ablative material would increase in roughness therefore the drag increment was changed to 0.02 and remained at this level for the rest of the simulated flight.

In the interest of flight safety the major aircraft systems had to function normally for the flight to proceed to the planned speed. Malfunctions possible during the flight dictated that a preplanned alternate flight be flown. As seen in the flight plan, the maximum speed was limited to 5,400 fps in case of failure of any of the dampers, attitude indication, ball nose . . . and/or external propellant flow failure at certain times. The ground rule for a failure of the engine to light on the first attempt after launch on previous tank flights had been to immediately eject the external tanks and fly an alternate profile. The main concern was that the maximum allowable dynamic pressure of 1,000 psf would be exceeded during the rotation (due to altitude loss) and that the dynamic pressure at the planned tank ejection would be too high for good separation characteristics. The possibility that a good mission could be lost to this cause was very real because delayed engine lights had occurred on six X-15 flights. For this flight the ground rule was changed to allow one restart attempt. The change was based in the following factors:

1. Increased confidence from previous flights in handling qualities at the planned [alpha] with the tanks on.
2. Simulator studies showed that the rotation could be performed at less than 1,000 psf dynamic pressure, particularly if the pilot flew at 2 degrees higher [alpha] and limited normal acceleration to 2.4 g. In addition, reduction of the throttle could be used as a positive method to keep the dynamic pressure less than 1,000 psf if necessary.
3. Simulator studies also indicated that if the dynamic pressure was too high at the time of planned tank ejection, no

detrimental effects resulted if the ejection was delayed until the dynamic pressure (which would be decreasing at this time) reached the desired value.

By flight date all the various alternate procedures were well known to the pilot after having practiced on the simulator for 35 hours. Fortunately, it did not become necessary to utilize an alternate profile since the flight proceeded basically according to plan.

The launch transients were very mild with a bank angle excursion of 14 degrees. During the rotation the pilot had good control of the aircraft and increased the angle of attack to 15 degrees and felt the onset of buffet. The remainder of the rotation to the planned pitch angle was made at 12 to 13 degrees angle of attack. During this period the roll control was excellent and the bank angle did not deviate more than 8 degrees. The maximum dynamic pressure experienced during the rotation was 560 psf, close to the 540 psf observed on the simulator. The planned pitch angle of 35 degrees was reached in 38 seconds and was maintained within plus/minus one degree.

The external tanks were ejected 67.4 seconds after launch. Tank separation was satisfactory, however, the pilot felt the ejection was "harder" than the last one he had experienced (Flight No. 2-50-89). The longitudinal trim change to the aircraft was from 4.2 to -2 degrees angle of attack. The external tank recovery system performed satisfactorily and the tanks were recovered in repairable condition.

After tank ejection the planned 2 degree angle of attack was maintained within +1 degree. As the aircraft came level at an indicated altitude of 99,000 feet, the pilot increased the angle of attack to 6 degrees to maintain zero rate of climb. During this task the pilot reported that the pitch control was very sensitive and it was difficult to hold a constant angle of attack.

The pilot reported shutting down the engine at 6,500 fps; however, the final radar data analysis revealed the maximum velocity to be 6,630 fps. The total engine burn time was 141.4 seconds, which compared favorably with the 141 seconds planned. However, the aircraft had achieved a velocity, which was 130 fps faster than that of the simulator during this time. . . . The ability of the ablative material to protect the aircraft structure from the high aerodynamic heating was considered good except in the area of the dummy ramjet where the heating rates were significantly higher than predicted. Considerable heat damage occurred on the dummy ramjet and the ramjet pylon. The ramjet instrumentation ceased approximately 25 seconds after engine shutdown indicating that a burn through of the ramjet/pylon structure had occurred. Shortly thereafter the heat propagated upward into the lower aft fuselage area causing the engine hydrogen peroxide hot light to illuminate in the cockpit. Ground control, assuming a genuine overheat condition, requested the pilot to jettison the remaining engine peroxide. The high heat in the aft fuselage area also caused a failure of a helium control gas line allowing not only the normal helium source gas to escape, but also the emergency jettison control gas supply as well (because of the failure of a check valve). Thus, the remaining residual propellants could not be jettisoned. The aircraft was an estimated 1,500 pounds heavier than normal at landing but the landing was accomplished without incident.

The heat in the ramjet pylon area became high enough to ignite 3 of the 4 explosive bolts retaining the ramjet to the pylon at some time during the flight. As the pilot was performing a turn to downwind in the landing pattern, the one remaining bolt failed structurally and the ramjet separated from the aircraft. The pilot did not feel the ramjet separate. Since the landing chase aircraft had not yet joined up, the pilot was not aware that the unit had separated. The position of the aircraft at the time of separation was established by radar data and the most likely trajectory estimated. A ground search party discovered the ramjet impact point on the Edwards AFB bombing range. Although it had been damaged by impact, it was returned for study of the heat damage that had occurred.

The unprotected right-hand windshield was, as anticipated, partially covered with ablation products. With the pilot's visibility being restricted (the left window was still covered by the eyelid) his guidance to the high key position was based on radar vectors from ground control. The eyelid was opened at approximately 1.6 Mach number as the aircraft was over Rogers Lake and the visibility out this window was good.

Ablative Performance

The ability of the ablative coating to protect the aircraft structure from high temperature was better than expected; the maximum temperatures recorded on flight 2-53-97 were lower than post-flight calculations based on the actual flight conditions encountered.

The condition of the ablative material after flight 2-53-97 was considered good.

The large amount of charring of the speed brakes was due to the high heat, which resulted from extending the speed brakes to 35 degrees shortly after maximum velocity was reached.

Heating in the Ramjet Area

The severe structural damage to the dummy ramjet and pylon during the flight to Mach 6.7 was the result of local aerodynamic heating due to shock impingement and flow interference effects.

The most severe melting damage occurred near the bottom of the ramjet pylon where shock waves generated by the ramjet spike tip, spike flare, cowl lip and bottom pressure probe were assumed to have intersected. A postflight thermal analysis of the heating in this area was made using the recorded temperature from the thermocouple located at the leading edge of the pylon and the observed heat damage as a guide. The measured temperature indicated a low value (less than 0 [degrees] F) until approximately 145 seconds after launch when a rapid rise in temperature occurred, indicating that the ablative material had burned through. The recorded temperature was increasing rapidly when the thermocouple wiring was severed by heat. A thermal analysis match of the ablator burn through time of 145 seconds was obtained when the undisturbed heat-transfer coefficient was increased by a significant factor. This analysis showed that the temperature was sufficient to result in the melting damage of the Inconel X (melting temperature approximately 2,600 degrees F) pylon structure.

Two areas of high heating due to flow interference were the pylon/fuselage junction and the ramjet cowl lip. A reasonable temperature profile for the indicated damage was also obtained by increasing the undisturbed heat-transfer coefficient by a significant factor. The calculated temperature due to interference heating of the ramjet cowl lip exceeded the melting point of the 4130 steel (2,800 degrees F) for a short time causing the melting damage.

Conclusions

The redesigned propellant system, with external tanks to contain additional propellants for increased performance capability of the X-15A-2, was brought to maturity through ground test stand development and flight test. Adequate tank separation characteristics were demonstrated and the designed recoverable/refurbishable concept of the external tanks proven.

A satisfactory room-temperature cure ablative material was selected from several candidate materials through wind tunnel arc tests and through flight test in small quantities on the X-15 aircraft. With the limited test result from 2 flights with a full coating on the aircraft it appeared that a satisfactory ablative material had been developed to protect the aircraft structure from the high temperature associated with flight of the aircraft outside its original design envelope.

A real-time analog simulation of temperature resulting from aerodynamic heating was developed. This simulation was used in conjunction with the six degree of freedom simulation of the aircraft to plan flights in which temperature at particular locations were one of the constraints of the desired flight. This combined simulation was also utilized during pilot training for the flight to enable the pilot to become aware of the effect of off-design flight conditions on the resulting temperature.

During the course of the flight program of the modified X-15A-2 several in-flight failures occurred which dramatically demonstrated the effects of relatively minor changes in configuration of a vehicle operating in an environment where aerodynamic heating is significant. Thermal loads caused premature extension of one or more components of the modified landing gear system at high Mach number on three occasions. Increased heating resulting from shock impingement and flow interference from the dummy ramjet installation caused severe structural damage to the ramjet and pylon.

The type of problems encountered during the course of the envelope expansion program may well be expected on other vehicles operating in the speed regime where aerodynamic heating will be an appreciable factor. In addition to the continued demonstration of piloted landing of an unpowered-low L/D vehicle, other techniques developed during the program are applicable to orbital lifting re-entry vehicles: application of and flight with an ablative coating, protection of a canopy window with a pilot-actuated covering, and development of an extendable pitot tube as an airspeed source for the terminal landing maneuver.

SELECTED READING

Books

Allen, Richard Sanders, *Revolution in the Sky,* Stephen Greene Press, 1967

Blue, Allan G., *The B-24 Liberator,* Charles Scribner's Sons, 1975

Borchers, Paul F., James A. Franklin, and Jay W. Fletcher, *Flight Research at Ames: Forty-Seven Years of Development and Validation of Aeronautical Technology,* NASA/SP-1998-3300

Bowers, Peter M., *Boeing Aircraft Since 1916,* Putnam/Aero, 1966

——— *Boeing B-29 Superfortress: Warbird Tech Vol. 14,* Specialty Press, 1999

Dabrowski, Hans-Peter, *Mistel: The Piggy-Back Aircraft of the Luftwaffe,* Schiffer, 1994

Green, William, *Warplanes of the Third Reich,* Doubleday, 1970

Jenkins, Dennis R., *Hypersonics Before the Shuttle: A Concise History of the X-15 Research Airplane,* NASA Publication SP-2000-4518, 2000

——— *Magnesium Overcast: The Story of the Convair B-36,* Specialty Press, 2001

Jenkins, Dennis R., and Tony Landis, *North American XB-70A Valkyrie: Warbird Tech Vol. 34,* Specialty Press, 2002

Johnsen, Frederick A., *Lockheed C-141 Starlifter: Warbird Tech Vol. 39,* Specialty Press, 2005

Leary, William M., *We Freeze to Please: A History of NASA's Icing Research Tunnel and the Quest for Flight Safety,* NASA SP-2002-4226, 2002

Machat, Mike, *World's Fastest Four-Engine Piston-Powered Aircraft,* Specialty Press, 2011

Saltzman, Edwin J., and Theodore G. Ayers, *Selected Examples of NACA/NASA Supersonic Flight Research,* NASA SP-513

Spearman, Arthur Dunning, *John Joseph Montgomery: Father of Basic Flying,* University of Santa Clara, 1967

Periodicals

"Exploring the Trans-sonic Flight Region," *The Martin Star*, June 1946

"For Probing the Transonic," *Aviation,* June 1947

"Laboratories in the Sky," *General Electric Review Staff Report*, May 1954

"Like No Goose You Ever Saw," *Air Progress*, February 1967

Meckley, W. O., "Flight Testing Gas Turbines," *Aviation Week*, 11 August 1947

"'Petulant Porpoise' Tests Hulls," *Aviation Week*, 12 July 1948

Story, Roger W., "They called them Flying Laboratories," *American Aviation Historical Society Journal*, Fall 2001

Story, Roger W., "The GE Schenectady Flight Test Center (its origins and early history)," *American Aviation Historical Society Journal*, Fall 2002

"Strangest Aircraft Ends Mission Here," *Desert Wings*, 13 January 1961

"Stump Jumper," *The Martin Star*, March 1947

"Three Musketeers of the Clouds," The Martin Star, June 1946

Reports

Armstrong, Johnny G., "Flight Planning and Conduct of the X-24A Lifting Body Flight Test Program," Air Force Flight Test Center, August 1972

Nelson, B. D., et al., "Investigation of a Micro-Fighter/Airborne Aircraft Carrier Concept Summary," Tactical Combat Aircraft Programs, The Boeing Aerospace Company, AFFDL TR 73-93 (Vol II), Air Force Flight Dynamics Laboratory, Air Force Systems Command, September 1973

Reynolds, Philip A., et al., "Capability of the Total In-Flight Simulator (TIFS)," Cornell Aeronautical Laboratory, Incorporated, February 1972

ENDNOTES

Chapter 1
1. Arthur Dunning Spearman, *John Joseph Montgomery: Father of Basic Flying*, University of Santa Clara, 1967.
2. Harlan D. Fowler (spelled Harland in the article byline), "The Fowler Variable Area Wing," *Aviation,* 21 July 1928.
3. Ibid.
4. Robert McLarren, "NACA Research Ends Ice Hazard," *Aviation Week,* 22 December 1947.
5. Ibid.
6. Kenneth S. Kleinknecht, "Flight Investigation of the Heat Requirements for Ice Prevention on Aircraft Windshields," *NACA Research Memorandum No. E7G28*, 5 September 1947.
7. Ibid.

Chapter 2
8. H. Heitman, translator; original author unknown, *The Planning and Development of Bombs for the German Air Force, 1925–1945*, Numbered USAF Study 192, AFHRA,1955.
9. Ibid.
10. Ibid.
11. Ibid.
12. Ibid.
13. Ibid.
14. Rene J. Francillon, *Japanese Aircraft of the Pacific War,* Naval Institute Press, Annapolis, Maryland, 1988.
15. Ibid.
16. Ibid.

17. *Ibid.*
18. *Anti-Suicide Action Summary*, CominCh P-0011, United States Fleet, Headquarters of the Commander in Chief, Navy Department, Washington 25, D.C., 31 August 1945.
19. *Ibid.*
20. *Ibid.*
21. *ibid.*
22. *Ibid.*
23. William Green, *Warplanes of the Third Reich,* Doubleday, Garden City, New York, 1970.
24. *Ibid.*
25. Typed note by Glen Edwards in his photo album in the AFTC history office collection.
26. Message, To: Commanding Officer, 1st XAS Squadron, From: Headquarters, 1st Experimental Guided Missiles Group, Eglin Field, Florida. Subject: Standard Operating Procedures for Air Launching, 9 October 1947.
27. *Ibid.*
28. *Ibid.*
29. *TSENG Report No. 3, formerly Report TSEST-A3, Characteristics of Experimental Airplanes (& Missiles) (Short Title "Green Book")*, 9th Edition, Air Materiel Command, 1948.
30. C. E. Anderson, "Aircraft Wingtip Coupling Experiments," Proceedings of the 23rd Annual Society of Experimental Test Pilots Symposium, 1979.
31. *Ibid.*
32. *Ibid.*
33. *Ibid.*
34. *Ibid.*
35. *Ibid.*
36. *Ibid.*
37. *Ibid.*
38. *Ibid.*
39. Summary Report, Project Tom-Tom, Submitted Under Contract No. AF33 (600)-23415, Convair; A Division of General Dynamics Corporation, Fort Worth, 30 December 1957.
40. *Ibid.*
41. *Ibid.*
42. *Ibid.*

Chapter 3
43. Allan G. Blue, *The B-24 Liberator,* Charles Scribner's Sons, New York, 1975.
44. *Preliminary Handbook, Part A, Operations and Flight Instructions: Models B-24C, B-24D, and B-24E Bombardment Airplanes, T.O. No. 01-5EC-1*, Revised 8-15-42, Consolidated Aircraft Corporation, San Diego, California.
45. "Rentschler Airport," *The Bee Hive*, United Aircraft Corp., Summer 1947.
46. *Ibid.*
47. Peter M. Bowers, *Boeing Aircraft Since 1916,* Aero Publishers, Fallbrook, California, 1966.
48. John F. Smith, "Flights for Proof," *United Aircraft Corporation's Beehive*, Summer 1956.
49. Scott A. Thompson, *Final Cut: The Post-War B-17 Flying Fortress: The Survivors,* Pictorial Histories Publishing Company, Missoula, Montana, 1990.
50. "Testbed...," *Boeing Magazine*, November 1948.
51. Report, *Propeller, Integral Gearbox Model 73EGB1 and Propeller, Variable Camber Model VC86260 Flight Test Report*, Bureau of Naval Weapons Contracts, NOw 64-0635-di (and) NOw 65-0533-d, 30 June 1966.
52. *Ibid.*
53. *Ibid.*
54. Scott A. Thompson, *Final Cut: The Post-War B-17 Flying Fortress: The Survivors,* Pictorial Histories Publishing Company, Missoula, Montana, 1990.
55. Ken Chilstrom and Penn Leary, editors, *Test Flying at Old Wright Field,* Winchester House Publishers, Omaha, Nebraska, 1991.
56. John F. Smith, "Flights for Proof," *United Aircraft Corporation's Beehive*, Summer 1956.
57. *Ibid.*
58. *Ibid.*
59. *Ibid.*
60. *Ibid.*
61. *Ibid.*
62. "Exploring the Trans-sonic Flight Region," *The Martin Star*, June 1946
63. Charles R. Anderson, "The TF-39 Flying Testbed," Proceedings of the 11th Annual Society of Experimental Test Pilots Symposium, 1967.
64. *Ibid.*
65. *Ibid.*
66. *Ibid.*
67. *Ibid.*

Chapter 4
68. E-mail message, Subj: Re: GE testbeds, 28 March 2017, from Jason Chapman, GE Aviation, US.
69. *Ibid.*
70. *Ibid.*
71. *Ibid.*
72. Robert C. Kohl, *Performance and Operational Studies of a Full-Scale Jet Engine Thrust Reverser,* NACA Technical Note 3665, Lewis Flight Propulsion Laboratory, Cleveland, Ohio, April 1956.
73. *Ibid.*
74. *Ibid.*
75. Peter M. Bowers, *Boeing Aircraft Since 1916,* Aero Publishers, Fallbrook, California, 1966.
76. *Ibid.*
77. *Flight-Determined Aerodynamic Properties of a Jet-Augmented Auxiliary-Flap, Direct-Lift-Control System Including Correlation with Wind-Tunnel Results,* NASA Technical Note TN D-5128, July 1969.
78. *Ibid.*
79. *ARTB: Advanced Radar Testbed Electronic Countermeasures,* Lockheed Aeronautical Systems Company, Marietta, Georgia (Undated).

Chapter 5
80. Kenneth H. Sullivan and Larry Milberry, *Power: The Pratt & Whitney Canada Story,* Vol. 1, 1928–1988, published by Pratt & Whitney Canada, 1989 and 2013.
81. *Ibid.*
82. *Ibid.*
83. *Ibid.*
84. *Ibid.*
85. *Ibid.*
86. *Ibid.*
87. *Power: The Pratt & Whitney Canada Story,* Vol. 2, 1989–2013, published by Pratt & Whitney Canada, 2013.
88. Kenneth H. Sullivan and Larry Milberry, *Power: The Pratt & Whitney Canada Story,* Vol. 1, 1928–1988, published by Pratt & Whitney Canada, 1989 and 2013.
89. *Ibid.*

Chapter 6
90. "Model Flies at High Speed," *Naval Aviation News,* August 1947.
91. Walter C. Williams, *et al, General Handling-Qualities Results Obtained During Acceptance Flight Tests of the Bell XS-1 Airplane,* NACA Research Memorandum RM L8A09, 19 April 1948.
92. Peter M. Bowers, *Boeing B-29 Superfortress,* Warbird Tech Series, Volume 14, Specialty Press, North Branch, Minnesota, 1999.
93. "Bell X-1A Research Airplane Crashes," *History of the Air Force Flight Test Center,* Volume I, 1 July to 31 December 1955.
94. *Ibid.*
95. *Ibid.*

Chapter 7
96. *System Development Plan, X-15 Research Aircraft, Supporting Research System Number 447L,* ARDC, 22 March 1956.
97. Wendell H. Stillwell, *X-15 Research Results With a Selected Bibliography,* NASA SP-60, 1965.
98. Maj. Jerauld R. Gentry, USAF, "Lifting Body Flight Test," Proceedings of the 12th Annual Society of Experimental Test Pilots Symposium, 1968.
99. *Ibid.*
100. *Ibid.*
101. *Application of Fracture Mechanics and Half-Cycle Method to the Prediction of Fatigue Life of B-52 Aircraft Pylon Components*, NASA Technical Memorandum TM-88277, 1989.
102. William L. Ko and Lawrence S. Schuster, *Stress Analyses of B-52 Pylon Hooks*, NASA Technical Memorandum 84924, 1985.
103. *Application of Fracture Mechanics and Half-Cycle Method to the Prediction of Fatigue Life of B-52 Aircraft Pylon Components*, NASA Technical Memorandum TM-88277, 1989.

Chapter 8
104. Ivy Hooks, David Homan, Paul Romere, "Aerodynamic Challenges of ALT," NASA Johnson Space Center.
105. *Comparison of 747 and C-5A as Shuttle Orbiter Carriers,* 4-25-74, author not noted, Space Division, Rockwell International.
106. Ivy Hooks, David Homan, Paul Romere, "Aerodynamic Challenges of ALT," NASA Johnson Space Center.
107. *Ibid.*
108. *Ibid.*
109. *Ibid.*
110. *Ibid.*
111. *Ibid.*

Chapter 9

112. *TSENG Report No. 3, formerly Report TSEST-A3, Characteristics of Experimental Airplanes (& Missiles) (Short Title "Green Book"),* 9th Edition, Air Materiel Command, 1948.
113. *Characteristics Summary, Drone...Q-2A,* 4 September 1956.
114. At the time of writing, the AQM-34 Firebee website developed by Rob de Bie at https://robdebie.home.xs4all.nl/aqm34.htm contains many details about the Ryan Firebee and its carrier aircraft.
115. Laurence R. Newcome, *Unmanned Aviation: A Brief History of Unmanned Aerial Vehicles,* American Institute of Aeronautics and Astronautics, Reston, Virginia, 2004.
116. *Ibid.*
117. Joseph Earl Dabney, *Herk: Hero of the Skies,* Copple House Books, Lakemont, Georgia, 1979.

Chapter 10

118. From the Scaled Composites website at www.scaled.com.
119. *Ibid.*
120. *Ibid.*

Chapter 11

121. William T. "Bill" Creech, Colonel, USAF (Ret.), *The 3rd Greatest Fighter Pilot,* Author House, Bloomington, Indiana, 2005.
122. *Ibid.*
123. Ken Chilstrom and Penn Leary, editors, *Test Flying at Old Wright Field,* Winchester House Publishers, Omaha, Nebraska, 1991.
124. At the time this was written, naval aviation historian Tommy H. Thomason had a thorough website with photos and details about the L-39s at: http://thanlont.blogspot.com/2011/04/bell-l-39-wing-sweep-evaluation.html
125. *Flight Measurements of the Stability, Control, and Stalling Characteristics of an Airplane Having a 35-Degree Sweptback Wing Without Slots and with 80-Percent-Span Slots and a Comparison with Wind-Tunnel Data,* NACA Technical Note 1743, November 1948.
126. *Ibid.*
127. *Ibid.*
128. *Ibid.*
129. *Flight Measurements of the Lateral and Directional Stability and Control Characteristics of an Airplane Having a 35-Degree Sweptback Wing With 40-Percent-Span Slots and a Comparison with Wind-Tunnel Data,* NACA Technical Note 1511, January 1948.
130. At the time this was written, naval aviation historian Tommy H. Thomason had a thorough website with photos and details about the L-39s at: http://thanlont.blogspot.com/2011/04/bell-l-39-wing-sweep-evaluation.html
131. *Ibid.*
132. Stanley N. Roscoe and Scott G. Hasler, *Flight By Periscope,* Illinois University at Urbana, 1952.
133. "Periscope Flight Testing," *History of the Air Force Flight Test Center,* Volume I, 1 July to 31 December 1955.
134. James R. Hansen, *Engineer in Charge: A History of the Langley Aeronautical Laboratory, 1917–1958,* NASA, Washington, D.C., 1987.
135. *Flight Investigation of a Supersonic Propeller on a Propeller Research Vehicle at Mach Numbers to 1.01,* NACA Research Memorandum RM L57E20, 10 July 1957.
136. *Ibid.*
137. *Ibid.*
138. *Ibid.*
139. *Flight Performance of a Transonic Turbine-Driven Propeller Designed for Minimum Noise,* NASA Memorandum 4-19-59L, May 1959.
140. "Republic Fighter Tests Turboprop," *Western Aviation,* September 1955.
141. *Low-Speed Longitudinal Stability Characteristics of a 1/6-Scale Model of the Republic XF-84H Airplane With the propeller Operating,* NACA Research Memorandum RM SL53F26, June 1953.
142. "XF-84H Flight Test Program," *History of the Air Force Flight Test Center,* Volume I, 1 Jan to 30 June 1955.
143. "Holtoner Semiannual Summary Letter," 10 Aug 55, excerpted in *History of the Air Force Flight Test Center,* Volume I, 1 Jan to 30 June 1955.
144. *Ibid.*
145. *Ibid.*
146. "XF-84H Phase I Test," *History of the Air Force Flight Test Center,* Volume I, 1 July to 31 December 1955.
147. "Turbojet Aircraft Noise Survey," Project Assignment Directive Number 55-39, Headquarters, Air Force Flight Test Center, 1 August 1955, *History of the Air Force Flight Test Center,* Volume II, 1 July to 31 December 1955.
148. Letter, Carl A. Bellinger to Frederick A. Johnsen, 11 December 1975, Subject: F-84.
149. Marcelle Size Knaack, *Encyclopedia of U.S. Air Force Aircraft and Missile Systems, Volume II, Post-World War II Bombers, 1945–1973,* Office of Air Force History, Washington, D.C., 1988.
150. *Ibid.*
151. "Boundary Layer Control Tests," *History of the Air Force Flight Test Center,* Volume I, 1 July to 31 December 1955.
152. Fred Anderson, *Northrop: An Aeronautical History,* Northrop Corporation, Los Angeles, California, 1976.
153. "Final Report, LFC Aircraft Design Data Laminar Flow Control Demonstration Program," Northrop, June 1967.
154. Fred Anderson, *Northrop: An Aeronautical History,* Northrop Corporation, Los Angeles, California, 1976.
155. *Ibid.*
156. "X-21A Category I Tests (659A)," *History of the Air Force Flight Test Center,* Volume I, 1 July to 31 December 1963.
157. Joseph W. Pawlowski, *et al.,* "Origins and Overview of the Shaped Sonic Boom Demonstration Program," American Institute of Aeronautics and Astronautics.
158. *Ibid.*
159. *Ibid.*
160. Dr. Alan W. Wilhite and Dr. Robert J. Shaw, "An Overview of NASA's High-Speed Research Program," ICAS 2000 Congress.
161. *AFTI/F-16 Automated Maneuvering Attack System Test Reports/Special Technologies and Outlook,* National Aerospace and Electronics Conference, May 19–23, 1986 (published by General Dynamics 11 July 1986).
162. *Ibid.*

Chapter 12

163. *Survey on Seaplane Hydro-Ski Design Technology, Phase 1: Qualitative Study,* Edo Corp., College Point, New York (P.A. Pepper and L. Kaplan), for Office of Naval Research and Naval Air Systems Command, 23 December 1966.
164. *Ibid.*
165. *Ibid.*
166. *Ibid.*
167. *Full-Scale Demonstration of Vertical Float Sea-Stabilization Concept, Phase I Report,* General Dynamics/Convair, San Diego, California August 1963.
168. "The 'Slip-Wing' Fighter," *Flight, The Pictorial Flying Review,* May 1941.
169. "Russian Concept of A-Powered 6,000–12,000 mph Transport of Future," *American Aviation,* 13 February 1956.
170. "Navy Emphasizes Safety of its Pilots in Future High Speed Jet or Rocket Plane," *Naval Aviation News,* December 1946.
171. Photo caption, NACA 17743, 1-27-47 (from NARA).
172. "F-80 Flies on Ramjet Power Alone," *Aviation Week,* 8 November 1948.
173. "Navy Flies First Ram-Jet Pilotless Aircraft," *Naval Aviation News,* January 1948.
174. Johnny G. Armstrong, *Flight Planning and Conduct of the X-15A-2 Envelope Expansion Program,* Air Force Flight Test Center (AFFTC) Technology Document No. 69-4, July 1969.
175. *Air-Launched Balloon System,* Air Force Flight Test Center, AFFTC-TR-77-42, Final Report, February 1978.
176. *Ibid.*
177. *Ibid.*
178. "Powered by the Propulsion Technology of the 1990s," McDonnell Douglas news release photo caption, July 1987.
179. Rene J. Francillon, *McDonnell Douglas Aircraft Since 1920: Volume II,* Naval Institute Press, Annapolis, Maryland, 1990.
180. Edwin J. Graber, *Overview of NASA PTA Propfan Flight Test Program,* NASA Lewis Research Center, Cleveland, Ohio, 1 February 1990.
181. *Ibid.*
182. *Ibid.*
183. *Ibid.*
184. *Ibid.*
185. "747 Tanker Feasibility Study," News Release 027.72, U.S. Air Force, 18 February 1972.
186. *Investigation of a Micro-Fighter/Airborne Aircraft Carrier Concept,* Tactical Combat Aircraft Programs, The Boeing Aerospace Company, Air Force Flight Dynamics Laboratory, Air Force Systems Command, Wright-Patterson Air Force Base, Ohio, September 1973.
187. *Ibid.*
188. *Ibid.*
189. *Ibid.*
190. *Ibid.*
191. *Ibid.*
192. James Harry Pollitt, "The Case for Engine Flying Testbeds," Proceedings of the 15th Annual Society of Experimental Test Pilots Symposium, 1971.
193. *Ibid.*
194. *Ibid.*
195. *Ibid.*

INDEX

A
Abbotsford International Air Show, 116
Aeroproducts, 135, 137, 151
Air Canada, 81
Air Force Advanced Radar Testbed, 76, 77, 199
Air Force Aeronautical Systems Division, 76, 176
Air Force Air Research and Development Command, 96
Air Force Flight Dynamics Laboratory, 198, 200
Air Force Flight Test Center, 76, 89, 133, 136, 140, 172, 186–187, 198–200
Air Force Flight Test Museum, 77, 101
Air Force Geophysics Laboratory, 173
Air Mobility Command Museum, 77, 176
Aircraft Engine Research Laboratory, 45, 166
Airmotive, 77
All-American Engineering, 159
Allen, Paul G., 127, 198
Allied Signal, 60–61
Allison, 41, 52–53, 134–135, 142, 151, 174–175
American Airlines, 34, 40, 110
American Aviation magazine, 160
Ames Research Center, 149–150, 163
Anderson, Charles R., 27, 29, 64–65, 199–200
Anderson, Maj. Clarence E., 26
Armstrong Flight Research Center, 74, 144–145
Armstrong, Johnny, 172, 187
Army Air Force, 19, 20, 32, 45, 52, 53, 56, 77, 80, 84, 85, 117, 129, 167
Arnold, General Henry "Hap," 32–33, 51, 186
Avco Lycoming, 67–68, 150
Avro Canada, 79

B
Baka, 15–17
Balloons, 6–7, 172–173
Balls-8, 75, 96–97, 100, 102–103, 106, 108
Beaird, Henry, 136, 138
Beeler, D. D., 89
Bell Aircraft, 86, 131
Bellinger, Carl A., 138, 200
Betty Jo, 166
Betty, 15–17, 166
Billeter, Orion D., 103
Bisontennial, 149
Black Hills Aviation, 51
Blimps, 160
Boeing Field, 43, 47, 138
Boeing Company, 41–42, 45–49, 54–57, 60–62, 65, 69–72, 74, 77, 81–83, 85–87, 91, 110, 113, 129, 141, 143, 148–150, 163, 167, 173–174, 176, 178, 198–200
Breuhaus, Walt, 141
Bright, George L., 101, 151
British Aircraft Corporation, 73

Brown Field, 175
Bureau of Naval Weapons, 51, 199
Butchart, Stanley, 87, 89

C
Calspan, 52, 141, 145
Canadian Forces Base St. Hubert, 82
Cardenas, General Robert, 86
CFM International, 72
Chamberlin, Fred S., 50
Connecticut Aeronautical Historical Association, 51–52
Continental Motors, 173
Convair, 23–25, 30, 34–36, 45, 74–75, 142, 147, 151, 158, 169, 198–200
Crossfield, Scott, 86
Curtiss-Wright, 44–51, 54

D
Dana, Bill, 104
Davis-Monthan Air Force Base, 31, 77
Davis, Horace, 60
De Havilland Aircraft of Canada, 148
Defense Advanced Research Projects Agency, 148, 186
Digital Fly-By-Wire, 144
Dobbins Air Force Base, 175
Dover Air Force Base, 176
Drop Test Vehicle, 105
Dryden Flight Research Center, 74, 107, 113, 115, 144, 154, 186
Dulles International Airport, 74, 163

E
École Nationale D'Aéronautique du Québec, 80
Edo Corporation, 159
Edwards Air Force Base, 19, 23–24, 43, 50–51, 61, 74–77, 86, 89, 92, 96–98, 100–105, 107–108, 110–112, 115, 127, 189
Edwards, Captain Glen, 18, 19
Eglin Field, 18, 199
Everest, Lt. Col. Frank K., 90, 137
Evergreen Aviation Technologies, 72
Experimental Guided Missiles Group, 18, 199

F
Fairchild Air Force Base, 24
Farnborough International Airshow, 94
Fay, Charles, 161
Federal Aviation Administration, 51, 174
Fertile Myrtle, 86
FICON, 23–25, 30, 121, 176, 186
Flight magazine, 160
Floatplanes, 78, 158
Ford Motor Company, 34, 35, 36, 40, 41
Fowler, Harlan D., 7–8, 142, 198
French Air Force, 59, 168

Fullerton, Gordon, 75, 112
Fulton, Capt. Fitzhugh L. Jr., 91, 111
Furtek, Lt. (jg) A. J., 160–161

G
Garrett AiResearch, 61
GE Aviation, 34, 72, 199
General Dynamics, 154, 199–200
General Electric Edwards Flight Test Center, 65
General Electric, 32–36, 40–45, 51, 53–55, 61–72, 128–129, 131, 153, 170, 173–175, 186, 198–199
Gentry, Maj. Jerauld, 103, 199
George Air Force Base, 69
German Luftwaffe, 11
Goodlin, Chalmers, 161–162
Grimes Flying Lab Foundation, 73
Grossmith, Seth, 150
Guggenheim School of Aeronautics, 7
Guidry, Skip, 111
Gulfstream Aerospace, 156

H
Haise, Fred, 112
Hamilton Standard, 6–7, 45, 48
Hamilton, Frank, 45, 48
Harrigan, Lt. D. Ward, 11
Haven, Gil, 54
Hendrix, Lin, 138
Hill Air Force Base, 31
Hohmann, Ben, 26
Holloman Air Force Base, 94
Holtoner, Brig. Gen. J. Stanley, 136–137, 200
Honeywell, 60–61
Horton, Vic, 111
Hughes Aircraft Company, 77, 116, 145
Hughes, Howard, 68, 77
Huxman, Joseph D., 103

I
Innis, Bob, 150
Internally Stowed Fighter, 21
International Aero Engines, 81
International Space Station, 102

J
Jameson, David, 129
Japan Airlines, 72, 113
Jensel, Joseph, 89
Johnson, Kelly, 172, 199
Joint Base Elmendorf-Richardson, 73
Jones, Bruce, 151

K
Kadena Air Base, 122
Kelly Air Force Base, 42
Kennedy Space Center, 74–75, 108–110, 114
Kern County Airport, 137
Kincheloe, Capt. Iven C., 91, 94

King, Jay L., 82–83, 103
Kivette, Lt. (jg) Frederick N., 11–12
Knight, Major William J., 124–127, 172, 187
Kolb, Arnold, 51
Krier, Gary, 144

L
La Guardia Airport, 34
Lambert Field, 134
Lambert, First Sgt. Lawrence, 160–161
Langley Field, 133
Langley Memorial Aeronautical Laboratory, 133, 136, 147
Levier, Tony, 167
Liberty Belle, 52
Lockheed-California, 175
Lockheed-Georgia, 122, 175
Lucas Aviation, 71

M
MacArthur Airport, 56
Maia, 8, 160
Maloney, Dan, 6–7
Marquardt Aircraft, 167
Marquardt, Roy, 167
Martin Aviation, 56–60, 74, 77, 147, 159, 160, 167
McDonnell Douglas, 135, 150, 173–174, 200
McKay, Jack, 87–88, 187
McMurtry, Tom, 111
Military Air Transport System, 72
Miller, Lt. Harold B., 12
Mojave Air & Space Port, 69–71, 75, 126
Mojave Desert, 61, 71, 86, 87, 94, 109, 112, 124, 140, 151, 157
Monstro, 20–21
Montgomery, John J., 6–8, 172–173, 198
Moss, Frank, 33
Muroc Air Force Base, 20–23, 85, 117, 160, 162
Murray, Major Arthur, 89

N
NACA Lewis Flight Propulsion Laboratory, 38, 39, 166, 169, 170
NASA Dryden Flight Research Center, 74, 107, 115, 144
NASA John H. Glenn Research Center at Lewis Field, 166
National Advisory Committee for Aeronautics, 9, 35, 37–40, 45, 72–73, 86–87, 89,90, 96, 130–136, 145, 147, 156, 158–159, 162–167, 186, 198–200
National Aeronautics and Space Administration, 37, 73–75, 86, 99–116, 126–127, 135, 143–144, 147–150, 153–158, 163–167, 169–172, 174–175, 186–187, 192, 198–200
National Air and Space Museum, 74, 86, 163
National Air Force Museum, 82
National Museum of Naval Aviation, 68
National Museum of the United States Air Force, 137, 143, 156, 166, 186
National Parachute Test Range, 173
Naval Air Station Lakehurst, 10, 160
Naval Air Station Moffett Field, 12
Naval Air Station North Island, 145
Naval Air Station Point Mugu, 145, 148, 167

Nessly, Ray, 59–60
New England Air Museum, 154
Norair, 140–141
North American Aviation, 43, 65, 68, 96–97, 105, 171, 188
Northrop Grumman, 52, 54, 73, 74, 80, 84, 85, 143, 145, 146, 148, 153, 154, 159, 162
Northrop-Hawthorne, 139
Northrop, 18, 34, 54, 67, 68, 82, 94, 120, 121, 139, 140, 143, 150, 160–162, 164, 166, 167
Northrop, John K., 139
Northwest Airlines, 61

O
Offutt Air Force Base, 123
Ohta, Ensign Mitsuo, 15
Operation Batty, 119
Operation Eisenhammer, 12, 14
Orbital ATK, 127
Orbital Sciences Corporation, 107
Orenda, 78–79

P
Pacific Air Museum, 145
Pan American World Airways, 34, 41, 70
Parkins, Wright A., 55
Peenemünde Army Research Center, 14, 18
Petulant Porpoise, 146–147, 198
Pima Air and Space Museum, 146–147
Pitcairn Airfield, 8
Pittsburgh Plate Glass, 9
Pollitt, James Harry, 145, 148, 167
Pratt & Whitney Canada, 78, 78, 80–83,
Pratt & Whitney, 18, 41, 42, 45–48, 51, 52, 54, 55, 61, 62, 64, 74, 127, 174, 175
Project MX-106, 27
Pueblo Weisbrod Aircraft Museum, 68

R
Radioplane Company, 120
Rawdon Brothers Aircraft, 173
Raytheon, 145
Rentschler Field, 55
Republic Aviation Corporation, 27, 133
Rockwell International, 143, 199
Rogers Dry Lake, 21, 75, 87, 100, 102–104, 109, 137, 172, 189
Rohr, 175
Royal Canadian Air Force, 78, 80, 82, 94
Rutan, Burt, 124

S
Safran Aircraft Engines, 72
Santa Clara, 6–7, 198
Scaled Composites, 3, 124, 126–127, 181, 200
Schiele, Major Joe S., 141
Schoch, Ed, 21
Seaplanes, 8, 158, 160
Selfridge Air Force Base, 170
Shuttle Carrier Aircraft, 75, 109–114

Sky Harbor Airport, 61
Smithsonian Institution, 86, 163
Snecma, 72
Society of Experimental Test Pilots, 64, 180, 199–200
Space Shuttle *Columbia*, 114
Space Shuttle *Discovery*, 115
Space Shuttle *Enterprise*, 109
Sperry Gyroscope Company, 55
Strategic Air Command, 30–31, 54, 65, 96–97, 122–123
Stratolaunch Systems, 116, 127
Sudderth, Robert, 141
Suicide attackers, 16

T
Talbot, C. G., 40
The Beast of Burbank, 77
The High and The Mighty One, 96
Thieblot Aircraft Company, 30
Thompson, Milt, 103, 199
Thurston Aircraft, 160
Tip Tow Program, 25, 27–30
Tom-Tom Program, 25, 29–31, 176, 199
Tonopah Army Airfield, 117
TWA, 61

U
Udvar-Hazy Center, 163
United Aircraft, 45, 51, 199
USS *Akron* (Navy Zeppelin), 11
USS *Kitty Hawk*, 149–150
USS *Los Angeles* (Navy Zeppelin), 10, 200
USS *Macon* (Navy Zeppelin), 11, 12
USS *Mannert L. Abele*, 16
USS *West Virginia*, 16

V
Vandenberg Air Force Base, 108
VISTA, 145, 186
Vogt, Dr. Richard, 26

W
Walker, Joseph A., 89
Westinghouse, 10, 20, 37–40, 45, 53–54, 57–59, 73, 94, 134, 151
Whale, 20, 145, 148
Whitcomb, Richard, 143
White, Major Robert M., 107, 124–127, 193
Wingless Wonder, 130
Woersching, Tom, 60
Woomera Range Complex, 94
Wright Aeronautical, 45, 49–50
Wright Field, 26, 53, 130, 160, 199–200
Wright-Patterson Air Force Base, 76, 200

Y
Yontan Airfield, 16
Young, Lt. Howard L., 11, 18, 20, 33

WORLD'S FASTEST SINGLE-ENGINE JET AIRCRAFT: Convair's F-106 Delta Dart Interceptor *by Col. Doug Barbier, USAF (Ret.)* This book provides an insightful and in-depth look at the sixth member of the Air Force "Century Series" family of supersonic fighters. From initial concept through early flight test and development and into operational service, every facet of the F-106's career is examined. Hardbound, 10 x 10 inches, 240 pages, 350 photos. *Item # SP237*

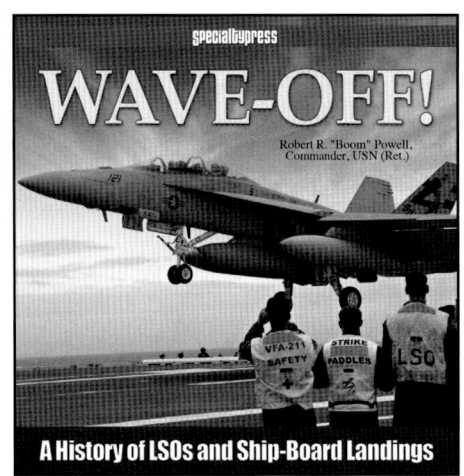

WAVE-OFF! A History of LSOs and Ship-Board Landings *by Tony Chong* This book tells the story of Landing Signal Officers from the first carrier operations in 1922 through World War II, the early jet era, Korea, Vietnam, and up to today's nuclear-powered leviathans. Also explained are naval aircraft and equipment development through the years; it covers both the faster and heavier aircraft and the changes in shipboard flight-deck systems. Diagrams showing the evolution of aircraft carrier deck design from World War I to the present are also included. Hardbound, 10 x 10, 192 pages, 188 b/w and 82 color photos. *Item # SP235*

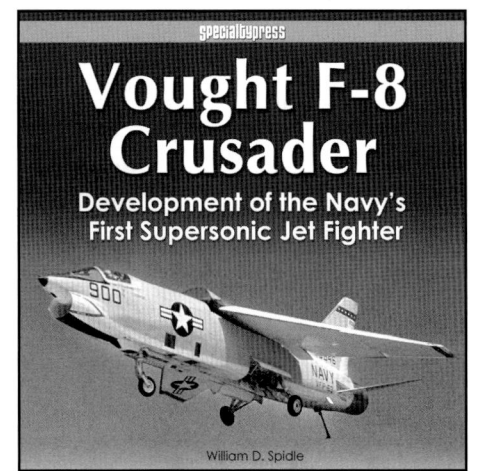

VOUGHT F-8 CRUSADER: Development of the Navy's First Supersonic Jet Fighter *by William D. Spidle* Detailed coverage of the Vought F-8 Crusader's design and development, production, record flights, the XF8U-3 Super Crusader, significant NASA variants, and foreign operators augments the story of this aircraft's development. In so doing, this book provides the vital "missing link" in the story of this legendary aircraft: the U.S. Navy's first supersonic jet fighter. Hardbound, 10 x 10 inches, 228 pages, 431 photos. *Item #SP242*

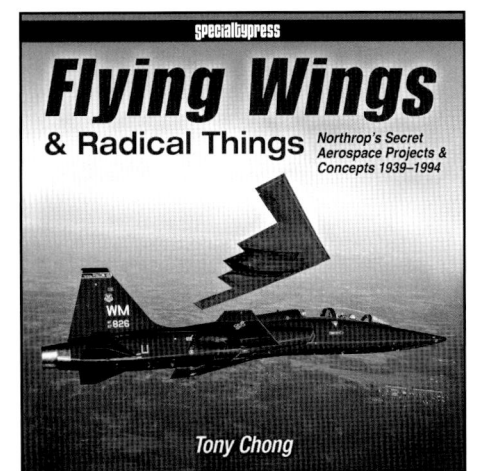

FLYING WINGS & RADICAL THINGS: Northrop's Secret Aerospace Projects & Concepts 1939–1994 *by Tony Chong* John K. "Jack" Northrop and the company he founded in 1939, Northrop Aircraft, Inc., will be forever linked with the giant futuristic Flying Wings of the 1940s. Here for the first time is the untold story of Northrop's rare, unique, and formerly super-secret aircraft and spacecraft of the future. Featuring stunning original factory artwork, technical drawings, and never-before-seen photographs, this book shows an amazing array of radical high-performance aircraft concepts from Jack Northrop and his team of brilliant and innovative engineers. Much of this material has only recently been declassified. Hardbound, 10 x 10, 192 pages, 361 b/w and 70 color photos. *Item # SP229*

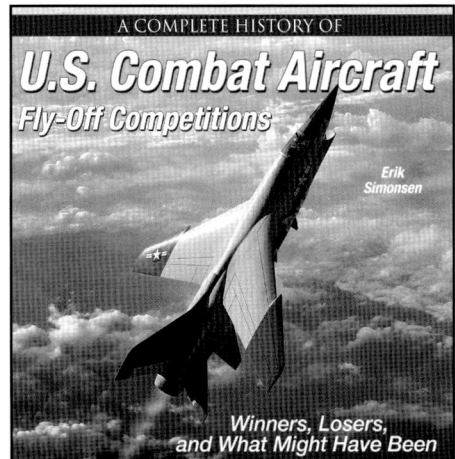

A COMPLETE HISTORY OF U.S. COMBAT AIRCRAFT FLY-OFF COMPETITIONS *by Erik Simonsen* Many advanced and now legendary aircraft have been designed, built, and flown in every generation of aviation development. Focusing on the Cold War era, this book shows readers how crucial fly-off competitions have been to the development of America's military air arsenal. This book explains in detail how fly-off competitions are conducted, and it shows what both competing aircraft designs looked like during their trials, and what the losing aircraft would have looked like in operational markings had it actually won. Hardbound, 10 x 10, 228 pages, 395 color and 156 b/w photos. *Item # SP227*

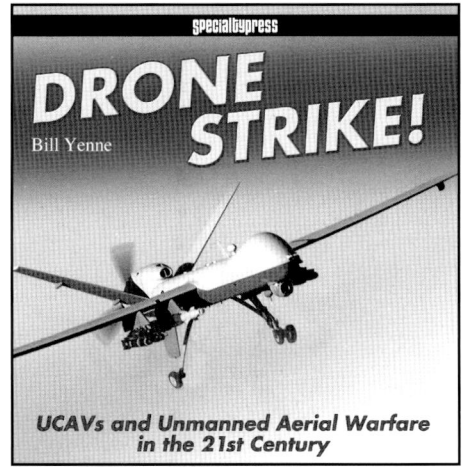

DRONE STRIKE!: UCAVs and Aerial Warfare in the 21st Century *by Bill Yenne* Takes you from the end of the 20th Century through today's latest technical wonders, covering such amazing unmanned aircraft capabilities as aerial refueling and landing aboard aircraft carriers even more accurately than manned aircraft. Hardbound, 10 x 10 inches, 160 pages, 300 b/w and color photos. *Item # SP238*

Specialty Press, 838 Lake Street South, Forest Lake, MN. Phone 800-895-4585 & 651-277-1400 Fax: 651-277-1203
www.specialtypress.com
Crécy Publishing Ltd., 1a Ringway Trading Estate, Shadowmoss Road, Manchester, M22 5LH, England. Phone: 44 161 499 0024 Fax: 44 161 499 0298
www.crecy.co.uk